SCIENCE THRILLER NOVELS BY AMY ROGERS:

PETROPLAGUE
REVERSION
THE HAN AGENT

SCIENCE IN THE NEIGHBORHOOD

How STEM professionals keep
Sacramento clean, dry, and moving

(Plus secrets of
how everyday things work!)

Dr. Amy Rogers

ScienceThrillers Media

SCIENCETHRILLERS MEDIA

Copyright ©2018 by Amy Rogers. All rights reserved. For information, contact Science Thrillers Media, P.O. Box 601392, Sacramento, CA 95860-1392. AmyRogers.com

Versions of these articles first appeared in *Inside Arden*, 2014-2018. All images obtained with rights from Shutterstock.

Publisher@ScienceThrillersMedia.com
www.ScienceThrillersMedia.com

Publisher's Cataloging-In-Publication Data
(Prepared by The Donohue Group, Inc.)

Names: Rogers, Amy, 1969-
Title: Science in the neighborhood : how STEM professionals keep Sacramento clean, dry, and moving : (plus secrets of how everyday things work!) / Dr. Amy Rogers.
Description: Sacramento, CA : ScienceThrillers Media, [2018] | "Versions of these articles first appeared in Inside Arden, 2014-2018." | Includes index.
Identifiers: ISBN 9781940419213 (paperback) | ISBN 9781940419220 (ebook)
Subjects: LCSH: Science--California--Sacramento--Popular works. | Engineering--California--Sacramento--Popular works. | Public utilities--California--Sacramento--Popular works. | Urban transportation--California--Sacramento--Popular works. | Ecology--California--Sacramento--Popular works. | BISAC: SCIENCE / Essays. | NATURE / Regional. | TECHNOLOGY & ENGINEERING / Social Aspects. | REFERENCE / Questions & Answers.
Classification: LCC Q162 .R64 2018 (print) | LCC Q162 (ebook) | DDC 509.79454--dc23
Library of Congress Control Number: 2018902552

To Jan and Duffy, who led me *Inside*

Table of Contents

Foreword ix

Transportation
Traffic signals 1
Seasonal gasoline 6
Distracted driving 10

Weather
Delta Breeze 17
Fog season 21
Managing flood risk 25
Earthquakes in Sacramento 37

Utilities
Wrangling electrons 43
Tap water 51
Wastewater treatment 59
Garbage 69

Ecology
American River salmon 83
Vernal pools 88
Save our elms 92
Great Backyard Bird Count 96
Nature Bowl 100
Mosquito and vector control 104

Environmental mercury 111
California Naturalist Program 115

Community
Science Cafés 121
Museum of Medical History 125
NorCal Herps 129
Cristo Rey High School 133
Powerhouse Science Center 137

Health
Defibrillators 143
Sports drink or water? 147
Blood donation and banking 151
Donating "good" bacteria can save lives 158
Seasonal allergies 163
Flu 167

At Home
Fruit ripening 173
The science of a perfect turkey 177
Glow sticks and fluorescence 181
Solar cooking 185
Swimming pool science 189

Index 193
About the Author 203

Foreword

Science is in your neighborhood.

Many people think of "science" as memorizing a bunch of facts, things like the periodic table or the stages of mitosis. While the efforts of scientists in the past have indeed given us many facts that must be learned (lest we be forced to repeat all their experiments!), science is much more than facts. It's an approach to learning based on questions and open-mindedness to the answers. It's a way of explaining and controlling the world around us. Once you start to look, you'll see that science—and her practical sister, engineering—are everywhere.

As a scientist, an educator, and a writer, I have a habit of seeing science in all things—in my home, my community, my region. But I have blind spots. Some everyday things I take for granted and I don't notice how they depend on good science and engineering. I'm sure you have the same problem. And it is a problem. Just because it's simple to turn on a tap, flush a toilet, or pump gasoline into your car, doesn't mean that the systems behind these conveniences are simple too. If we don't appreciate how complex it is to run a utility, then we will resent that it costs us money. If we don't understand where our waste goes, then we have no reason to be careful about what we throw away.

To answer "how does that work?" questions for myself and for the community, several years ago I started writing a "Science in the Neighborhood" column for the hyperlocal newspapers Inside Arden, Inside East Sacramento, and Inside

Land Park. It's been an extraordinary opportunity to meet local science experts and engineers. I've toured restricted facilities like the control center for our high-voltage power grid, and our regional blood bank. I've interviewed National Weather Service meteorologists, and the executive director of the Sacramento Area Flood Control Agency.

Over and over, I've learned one very important lesson: Things are more complicated than you think.

From my very first article about the engineering behind traffic signals at local intersections, I've been impressed by the sophistication required to make things work. Science, technology, engineering, and math professionals, many of them employees of our local governments or supra-government districts, apply their knowledge and skills behind the scenes. Without them, we'd be uncomfortable, sick, and isolated in the dark—or worse.

By collecting my columns into book form, I want to celebrate these hidden science heroes, and possibly inspire a young person to consider a STEM career. I'll also introduce you to citizen science projects you can join; unique local ecology you can visit; and places you can hang out with science-minded people. I'll explain ways to save someone else's life and how to protect your own while driving, fishing, or exercising. But most of all, with this book I want you to see, as I do, the importance of science in our homes and neighborhoods.

Note to *Inside* readers: I've updated and expanded many of these topics to include a greater depth of information than I originally had room for.

Amy Rogers, MD, PhD
Sacramento, California

@ScienceThriller
AmyRogers.com

Part One

Transportation

Red Light, Green Light: For Traffic Engineers, Timing is Everything

> Like every city in California, we have congestion problems on our roads. To get the best traffic flow possible on local streets, virtually every intersection with a traffic light is monitored and controlled. Not by random chance do you approach an empty intersection and the light turns green for you. Now I know why I never have to slam on the brakes for a yellow light: the engineers have the timing figured out.

Under cover of pre-dawn darkness, your platoon moves forward. All you want is to get through. Then it happens—you're in the Dilemma Zone, and your presence has been detected. The metal container shielding your body triggers eddy currents in a hidden wire, and now the ITS is taking control.

A scene from a military sci-fi video game?

No. Just the morning commute in Sacramento.

Fighting traffic feels like you against the world, but you're *not* alone. Engineers for the City and County of Sacramento Departments of Transportation are using advanced detection and communication systems, as well as old-fashioned mathematics, to help each automobile, bicycle, and pedestrian win the battle against congestion.

We all know the problem. Too many vehicles compete for space on Sacramento streets. In most cases, adding more lanes isn't feasible. So traffic engineers operate Intelligent Transportation Systems to make the most of the roads we have.

Traffic signals at intersections are the key to smooth traffic flow. Coordinated traffic signals allow a group of vehicles (called a *platoon*) to roll through one intersection after another, hitting green lights as often as possible.

How? In the simplest case—a one-way street, such as 16th Avenue downtown—the lights are programmed with a timing plan that calculates, based on distance and speed, how long it should take a platoon to travel from one intersection to the next. Other major commuter corridors are more complex, with two-way traffic plus many cross streets. Left and right turns, pedestrians, and cyclists all compete for time and space in the intersections. Giving a few seconds to one user takes from another, so trade-offs must be weighed across the entire system to synchronize progression. Light rail versus vehicles, main corridors versus side streets, and regional versus local concerns all must be considered. For example, to prioritize getting cars off highway 50 can lead to big jams at intersections on local streets such as Watt Avenue.

County Transportation Engineer Doug Maas has been balancing the needs of these different travelers for over 25 years. According to Maas, the solution to optimal traffic flow is to combine timing plans with intelligent sensors and a human touch. For example, during the morning and afternoon rush periods, traffic signals are told to favor cars moving in the dominant direction. During the holiday season, signals on Arden Way "know" to accommodate more vehicles visiting Arden Fair Mall.

But minute to minute, the number and kind of users at an intersection is unpredictable. That's where detectors come in.

Smart intersections sense big metal objects like cars and smaller ones like bicycles. Cameras, radar, and microwave detectors may be located over the street, on the mast arm that also holds the signal "head" (the red, yellow, green light). Embedded in the pavement are magnetometers and the most common kind of detector, inductive loops. (You can often see big looping lines in the pavement where these are buried.) These are loops of wire that constantly carry an electric current. When a metal object passes over the buried wire loop, the object's magnetic field disrupts the current.

When a detector "sees" a vehicle, it communicates with the intersection's brain, which is housed in a signal controller cabinet standing near one corner. The controller takes appropriate action. Usually this means giving the vehicle a green light. At an *actuated* intersection like this, green lights are never wasted on an empty lane, and when a car approaches, it quickly gets permission to pass.

Detectors relay their information via fiber optic or copper cables to nearby intersections to keep the whole corridor running smoothly. Data is also carried to the Traffic Operations Center.

At the county's Traffic Operations Center near Bradshaw and Kiefer, a bank of computers faces an entire wall of bright, high-resolution monitors. The screens are filled with live video of traffic at any of hundreds of connected intersections. Here, Maas and his team of engineers are traffic gods. From this remote location they can manually operate any signal lights on the network, and watch the effects on traffic in real time. This is critical during the morning and evening weekday commutes, when engineers are on duty to iron out wrinkles in the traffic flow. With a little ingenuity—adding a few seconds of green light here or red light there—Maas can usually clear unexpected back ups in two or three signal cycles.

Every timing decision demands a trade-off. As Maas says, traffic planners are "fighting for seconds at over-capacity intersections." If the system gives extra time to left turn traffic, other users—say, pedestrians—lose time to keep the whole corridor in sync.

But not all users are equal. When emergency vehicles are speeding to the rescue, they *need* green lights. Thus Sacramento Metropolitan Fire engines and ambulances are equipped with infrared strobe light emitters that trigger a "high priority pre-emption" of the signal timing at an intersection. As an emergency vehicle approaches, the signal controller changes the lights to give the responders a clear path. (Contrary to the urban myth, you can't fool the detector by flashing your headlights.)

I asked Maas about the notorious red light cameras at some intersections. He insists the traffic engineers set parameters for each signal to maximize safety and capacity. They don't make artificially short yellow lights in order to trap motorists and raise revenue for the Sacramento

County Sheriff's Department, which enforces the red light violations.

While no signal technology can eliminate the evening jams at Watt and Fair Oaks, watchful engineers shepherd the traffic flow as efficiently as possible. They confirm their data, too, with what Maas called "windshield factor"—driving routes themselves in real life and measuring total drive time and number of stops. Maas's department consistently beats national averages on the National Traffic Signal Report Card. The next time a green light stays lit just long enough for you to get through, you can bet it wasn't random luck. The ITS is on your side.

Summer blend is not a coffee: Seasonal gasoline

The refining of gasoline is possibly the most consequential science/engineering business that nobody understands. Sacramento has only a tiny local petroleum industry, and no refineries, so I haven't written in depth about the process. But here is a glimpse of some of the complexity behind the pump.

In late spring or early summer, the price of gasoline typically rises. Part of the reason is supply and demand: Americans consume more gas for travel in warm weather. Another reason is regulatory, based on science: summer blend gasoline costs more.

What is summer blend gas, and why do we use it?

First, some background. Gasoline is not a single pure thing like water. It's a blend of different hydrocarbons derived from crude oil. Refineries adjust the blend and include non-hydrocarbon additives to meet desired specifications for the gas. For example, depending on the "recipe" they can change the octane rating to produce the different grades of gasoline you see at the pump.

Another property of gasoline they can change with the "recipe" is vapor pressure. Vapor pressure is a measure of how much of a liquid spontaneously evaporates into the air. In a closed container, you could measure how much the vapor rising from a liquid adds to the air pressure. A volatile liquid produces a lot of vapor and has a high vapor pressure (examples: rubbing alcohol, nail polish remover). A liquid that doesn't evaporate much has a low vapor pressure (example: cooking oil). Intuitively you know that vapor pressure rises with temperature. Hot liquids evaporate faster than cold ones. If the temperature rises enough, a liquid will reach its boiling point—the point at which the vapor pressure is greater than the atmospheric pressure.

For gasoline to ignite properly in your car's engine, the vapor pressure (called the Reid vapor pressure or RVP) must not be too low. On the other hand, if the Reid vapor pressure is too high, the gas will evaporate. Evaporated gasoline is a nasty air pollutant. It also costs the consumer money in lost product. Therefore federal and state laws require refineries to adjust their gasoline blends to keep the RVP below a certain threshold.

That regulatory threshold varies with the seasons. In winter, low temperatures naturally reduce the vapor pressure of all liquids. That means refineries can blend their gasoline with components that have a greater tendency to evaporate. One such component is butane. Butane is volatile but it's also cheap and abundant. As long as it's cold outside, supplementing with butane is an economical way to produce more gasoline as cheaply as possible.

At warmer temperatures, winter blend is unacceptably volatile. On a hundred-degree day, the butane component would escape into the air. A different blend, one with a lower vapor pressure, is needed to minimize evaporation of the gasoline in summer. Therefore in early spring, refineries reset their facilities to produce summer blend gas. Summer blend is less volatile than winter blend. It's also a little more expensive, because refineries cannot blend in cheap butane as a supplement.

For Sacramento residents, the higher cost at the pump comes with one minor and one major benefit. Summer blend is slightly more energy dense (according to AAA, 1.7% more). This translates into slightly better gas mileage for summer travel. But the big reason to switch to summer blend is air quality.

From May to October, the Sacramento region is prone to periods of unhealthy smog and elevated ozone levels. In fact, we are in a "severe nonattainment area," meaning that ozone levels can badly exceed a federal eight-hour standard. Along with Los Angeles, Bakersfield, and Fresno, Sacramento is in the top ten most ozone-polluted cities in the US. Local geography is partly to blame: temperature inversion layers often form here in the Central Valley. An inversion is a weather condition when a thick layer of warm air lies on top of cooler air near the ground. This is not normal: typically, air gets *colder* as you go up into the sky. Under normal conditions, cooler (and therefore denser) air from up high sinks and naturally mixes any layers. This tends to disperse ground-level air pollutants. During an inversion, mixing doesn't happen. The warm air layer acts like a cap, and air quality at the ground becomes poor because pollutants are trapped and concentrated.

During sunny, warm conditions, evaporated gasoline reacts with oxygen to form ground-level ozone. Ozone usually peaks in the heat of the afternoon and early evening, then dissipates during the cooler nighttime. Because ground-level ozone pollution is such a problem in our state, the California Air Resources Board has set high standards for summer blend gasoline—significantly higher than the federal EPA requirements. The gasoline you buy in Sacramento is "reformulated," the cleanest burning, lowest vapor pressure gasoline on the market. The regulatory season in Sacramento is also one of the longest. EPA requires summer blend at retail nationwide from June 1 to September 15. In Sacramento, the switch to summer blend begins May 1 and continues until October 31.

Along with "Spare the Air" efforts to decrease pollution from the exhaust of cars, trucks, and buses, summer blend gasoline is a tool to improve our local air quality and our respiratory health.

If you're interested in petroleum, you might enjoy my science thriller novel PETROPLAGUE, about oil-eating bacteria that contaminate the fuel supply of Los Angeles and paralyze the city.

Your Brain on Phone: Distracted Driving is Driving Blind

The science of attention is getting lots of, well, attention these days. The rise of the mobile phone and the ubiquity of dashboard screens in new cars is creating dangerous conditions on the road. Think you're not distracted when you take that call? Cognitive science says you're wrong.

~~~~~~~~~~

Your kid didn't feel so good before school today. You're collaborating on an important work project with an out-of-town colleague. You need to stay connected, keep on top of things. So even though you're driving, when you hear that "ping" you glance at the text message on your cell phone.

In a fraction of a second, you go from sober to drunk. You're piloting 3,000-plus pounds of deadly missile, and at highway speed, you'll travel the length of a football field as if you were blindfolded.

Maybe *you* don't use your phone in a moving car, but a staggering number of drivers do. A 2013 survey found that 98% of drivers correctly think texting and driving is dangerous, but 43% of those surveyed read texts nevertheless.

Why do we do it?

Neuroscientists are grappling with this question. The answers lie deep in what it means to be human. Our desire for community, for relationship with others, is ancient and powerful. When we experience contact with another person, our brain rewards us by releasing a bit of the neurotransmitter dopamine. Dopamine gives a sense of well-being. In effect, answering a text is a small-scale form of the same pleasure people get from using cocaine. Dopamine is a big part of the reason why when it comes to distracted driving, many people say one thing but do another.

Nearly 70% of California drivers say they have been hit or nearly hit by a driver using a cell phone. In 2012, 3,328 Americans died in distraction-related collisions. Texting while driving increases your crash risk by *twenty-three times*. This exceeds the risk of driving under the influence of alcohol.

Not me, some people think. I'm a good driver and I'm good at multitasking.

Modern cognitive science tells us that multitasking is a myth. The attention centers of our brains can't do two things at once. Our brains actually juggle two tasks, alternating focus from one to the other. Each time the focus switches, things can get missed.

With devastating consequences. In his excellent book *A Deadly Wandering*, Northern California writer Matt Richtel tells the poignant story of a young Utah man who killed two rocket scientists when he crossed the center line while texting. The man's subsequent performance on neurologic tests of attention showed his ability to multitask was actually better than average, giving lie to the notion that being "good" at multitasking can protect you from distraction.

When we pay attention, different parts of the brain synchronize to each other, and separate tasks require separate "tuning of the frequency." For example, the part of our brain that receives input from the eyes communicates with the part of the brain that processes or interprets what we see. When we make a choice to attend to a text message, what neuroscientists call top-down activation, the neurons that read text become sensitive, alert specifically to the message even if there are other competing objects in the field of view.

Thus the price of attention to one thing is ignoring something else. To focus on a text message, your brain *reduces* the sensitivity of neurons that process glimpses of things that are *not* texts—things like a bicyclist on the road.

The result is called *inattention blindness*. A person can "see" something with the visual cortex part of the brain that's connected to the eyes, but never become aware of the object because the *processing* part of the brain is paying attention to something else.

A Harvard psychologist famously demonstrated inattention blindness with a widely circulated video (the-invisiblegorilla.com/videos). Six people wearing either white or black shirts are milling about with a basketball. Viewers are instructed to count how many times players wearing white pass the ball. Midway through the short video, a woman in a gorilla costume saunters into the screen, pauses to beat her chest, and walks away. When questioned after the video, half of viewers missed the gorilla. Their eyes were on it, but their conscious minds never perceived it.

Therefore the problem with cell phone use by drivers isn't taking their *eyes* off the road. It's taking their *minds* off.

Inattention blindness can result from attention to non-visual inputs, too. Recent data suggest that talking on a cell phone is equally distracting whether or not you're hands-free (as is required by California law.) If your brain is busy processing a conversation, to minimize distraction, the brain curtails visual processing. Talking to someone who isn't physically in the car can blind the driver to the unexpected, even if the driver's hands never leave the steering wheel. (Interestingly, having a conversation with someone in the passenger seat is much safer, possibly because social and physical cues mold the conversation to match what's happening on the road. For example, during a difficult merge the chatting momentarily stops.)

The brain is powerful but when it comes to attention, its capacity is far from infinite. When you're behind the wheel, pay attention only to your driving. Arriving safely to a hug from your loved one will give you all the dopamine you need.

Part Two

Weather

Delta Breeze is the coolest thing in town

I lived in St. Louis for twelve years and loved the spectacular spring and fall weather. What I didn't love was the unrelenting humidity in summer. As a rule, once you turned your air conditioner on, you'd be keeping it on 24 hours a day. Imagine my delight to discover that despite the heat in Sacramento, most summer nights I can open my windows. Where does our natural air conditioning come from?

In other parts of the country, sunset gives little relief from sweltering summer days. Air conditioners run all night long.

They don't have our Delta Breeze.

Sacramento lies ninety miles from the coast but thanks to physics and a quirk of geography, on many hot days we get natural air conditioning from the ocean. The Delta Breeze is a sea breeze that pushes cooler marine air into our backyards, and brings our nighttime temperatures much lower than they would be without it.

The Delta Breeze is created by temperature differences between the land and sea. In summer, the land of the Central Valley gets very hot. It heats the air, and just like in a hot air balloon, that hot air rises. As it rises, it creates a tiny drop in the air pressure at ground level.

Meanwhile on the San Francisco Bay, the surface temperature of the ocean stays about the same year round, day or night, usually in the mid-50s. This keeps the air cool as well as humid or foggy. This cool, moist air forms a blanket called a *marine layer* sandwiched between the ocean and warm, dry air higher up in the atmosphere. The marine layer keeps the Bay Area closer to the temperature of the ocean. It stretches inland and is trapped by the Coastal Ranges, where it forms a wall of cloud that cars going west on I-80 sometimes plunge into.

But there's a gap in the Coastal Ranges at the Carquinez Strait, a narrow bit of water spanned by the toll bridge just past Vallejo. Here the waters of the Sacramento and San Joaquin Rivers flow out of the Delta and into San Pablo Bay on their way to the Pacific Ocean. It's also where the marine layer sneaks through the mountains and blows into the Central Valley.

The breeze is driven by the tiny difference in air pressure (only hundredths of an inch of mercury) between the hot valley (where air rises) and the cool coast (where air stays low and heavy). It's enough to generate a wind of at least 14 mph, with gusts of 20-25 mph common in Sacramento. In Fairfield, closer to the Carquinez gap, gusts reach 35-40 mph. Once the breeze reaches the

Central Valley, it spreads out, turning northeast toward Sacramento and Marysville, and southeast to Stockton.

The same phenomenon is at work in the winter too, but the breeze, driven by the temperature difference between the valley and the ocean, is inconsequential. Winter highs in Sacramento are typically in the 50s, which is also the surface temperature of the ocean. So hardly any inland air movement results.

In summertime, the effects are profound. According to James Mathews, a meteorologist for the National Weather Service in Sacramento, "If it weren't for the Delta Breeze, we'd have a lot more hundred degree days." As proof, Mathews told me that on average, Sacramento experiences 22 such days per year, and Stockton, 18. Redding, which is cut off from the Delta Breeze by its elevation in the mountains, suffers through 40.

"The breeze is self-regulating," Mathews says. A very hot day in the valley causes a big temperature gradient with the coast, which drives a strong Delta Breeze. The breeze then cools the valley and decreases the temperature difference, especially the next day. "In 2013, the breeze gave us our greatest one day temperature drop ever, from a high of 110 degrees on July 4th to 86 on July 5th."

The Delta Breeze doesn't always come to the rescue. Forecasting whether the breeze will arrive on a given day is particularly important to the power companies. "SMUD needs to match electricity supply with demand," Mathews says. "If the breeze isn't coming, they have to buy power for air conditioning."

So why do we get relief some days but not others? The depth of the marine layer is the primary determinant of the breeze. The thicker the layer of marine air over the coast, the more cool air spills into the Central Valley. 2000 feet at the coast is the minimum depth required to get a breeze here. The location of the *Pacific High* is also important. The Pacific High is a mass of warm, dry, high-pressure air that typically forms around northern California in the summer. If the Pacific High forms offshore, the marine layer is deep and the Delta Breeze is strong. If the Pacific High moves inland, it squashes the marine layer and the breeze dies out.

Meteorologists also try to predict the *time* the Delta Breeze will arrive. According to Mathews, "The sea breeze will pick up at the coast around noon, and reach us usually around 4 PM." If it arrives before the hottest time of day (usually the 5:00 hour), the breeze reduces the daily high and decreases electricity demand for the whole evening.

Mathews and his fellow scientists study this airy phenomenon but they don't always get it right. "If you're in this business long enough, you realize that 'nature' doesn't read your forecasts."

Fog Season

I love fog, the way it cocoons me and dampens sound. Like the end of a rainbow, fog always seems to be little further ahead, magically receding away as I approach. They say the Central Valley used to get more of it.
Why?

This article ran in January.

Whether you think of it as a blanket, a shroud, or in Carl Sandburg's words, something that comes on little cat feet, fog season is here.

The Central Valley is famous for its thick fog known as tule (too-lee) fog. Named after a reed found in local marshes, tule fog can be incredibly dense and can reduce visibility to dangerously short distances. Such fogs have been responsible for horrendous accidents on California freeways, such as a 108-car pileup near Fresno in 2007. Tule fogs can last for days, turning the world gray and dim.

If you're a long-time resident of Sacramento, you may have the feeling that our fogs aren't quite what they used to be. You may be right. According to research published in 2014 (Baldocchi & Waller, Geophysical Research Letters), the number of winter "fog events" declined by 46 percent over the previous 32 winters. There is also tremendous variation in the amount of fog we get from year to year. Why?

The answers are drought, development, and climate change. Formation of fog, and especially a dense tule fog, requires particular conditions. The three main ingredients are wet ground, calm air, and cold nights. In Sacramento, we typically experience these three from November to March. Fog is basically a cloud that forms at ground level. It can begin to form any time after the sun goes down. It often thickens during the night, and can linger far into the next day.

According to meteorologist Bill Rasch of the National Weather Service, most of our local fog is a type called radiation fog. Radiation fog comes after rain, and it depends on a temperature difference between the ground and the air. After sunset the ground cools by giving off—or radiating—heat (hence the name "radiation fog"). The air immediately above also becomes cooler through its contact with the ground. If the air cools enough, its water vapor will condense into tiny airborne droplets. Collectively, those drops of water are what make fog. This is physically the same phenomenon as when water condenses on the outside of a glass of ice water (air near the glass is cooled to its dew point, and water vapor turns to liquid).

Cold air is denser than warm air, so the cool air formed at night near the ground tends to form a layer at the bottom (called an *inversion*). The thickness of the cold air layer determines how thick the fog is. This in turn depends on how much the air layers are mixing. Fog won't form when it's windy because the cool ground air gets stirred with warmer air above. If the air is perfectly calm, the fog layer will be relatively thin and tight against the ground. In a slight breeze, the fog layer will get thicker as the cold air at the ground moves around a little and chills a thicker layer of air.

Fog is particularly scarce during drought years because the ground is so dry. Moisture evaporating from the soil provides the humidity needed for fog to form. The best winters for fog have periodic storms followed by long periods of high pressure (dry days). In *very* wet years, a lack of clear, cold nights can diminish the number of foggy days even though moisture is abundant.

The decrease in fog is good for motorists but an alarming development for California farmers. Some of our state's most valuable crops—almonds, pistachios, cherries, apricots and peaches—require a winter dormant period brought on and maintained by colder temperatures. Fog has been an important part of the winter cooling mechanism for these trees. Without fog, and with temperatures rising globally, fruit and nut yields are threatened.

Radiation fog can be very patchy. It's affected by ground cover, local temperature differences, and soil saturation. For example, fog is more likely to form over a rice paddy than a parking lot. The Central Valley is also subject to another kind of fog. Rather than forming locally, it blows in from the ocean, through the Vacaville area and spills into the valley. This tule fog can fill the entire Central Valley from Redding to Bakersfield, and is clearly visible from space. The fog can last for days.

We say that fog "lifts." What really happens is daytime sun warms the earth, which heats the lower air and evaporates the lower part of the fog first. Ultimately fog "burns off" when the lower air has become warm enough to destroy the inversion (separate layers of cool and warm air), and the fog turns back into invisible water vapor. Or a change in the weather that brings wind can mix the inversion layers.

Why can Sacramento residents sometimes escape the fog by driving a few miles up highways 80 or 50? Meteorologist Rasch says it's because cold air sinks and warm air rises. "Fog forms where the coldest air falls down into the valley. In winter the air actually gets warmer as you go up into the foothills—until you go up a lot."

What if you find yourself in the perilous situation of driving in fog? The most important thing to do is slow down. With poor visibility, collisions can happen fast, and the visual distortion of the fog can ruin your natural sense of how fast you're going. Second, make your car visible by turning on low-beam headlights, which also triggers your taillights. If your car is equipped with fog lights, use them, but never use high beams in fog. The glare (reflection of light off the airborne water droplets) makes it harder for you to see the road ahead. If conditions get really bad, turn on your hazard lights and get as far off the road as possible. *Never* stop your car on the road itself.

It was dry when I built it: Managing flood risk in Sacramento

For river flooding, Sacramento is the most at-risk city in America. That risk will never go away, but an elaborate (and costly) regional system of levees and dams provides some protection. Could our flood control system withstand a Hurricane Katrina or Hurricane Harvey? Probably not, but read this three-part series to learn what is being done to keep us dry, and what the challenges are.

Levees

In the winter of 1861-62, ten to fifteen feet of early snow fell in the Sierra Nevada. Then a series of huge, warm Pacific storms melted that snow and dumped more than *five feet* of rain in Los Angeles. The result: a California megaflood. The entire Central Valley became an inland sea. Downtown Sacramento drowned under ten feet of water. The state's newly elected governor, Leland Stanford, traveled by boat to his January inauguration and the California legislature decamped to San Francisco. One-third of all the property in the state was destroyed. Sacramento remained underwater for months. (ScientificAmerican.com: atmospheric-rivers-california-megaflood)

Could it happen again?

It seems strange to worry about flooding during a time of drought, but when it comes to water in the American West, the historical pattern is either feast or famine.

The flood of 1862, which followed two decades of exceptionally dry weather, is a warning that the natural forces driving the water cycle in our region can surprise us with their ferocity—and overwhelm our defenses.

"The 1986 flood was a wake-up call," says Rick Johnson, Executive Director of the Sacramento Area Flood Control Agency (SAFCA), a joint powers agency formed in 1989 to address our vulnerability to catastrophic flooding. In the hundred years after the megaflood, Sacramento did a lot to protect itself. Downtown was raised by 10-15 feet; a system of levees was built to shield homes and farmland; Folsom Dam was designed to hold back storm water and melting snow.

And yet, in 1986 the system came perilously close to failure. "If we'd had another four to six more hours of rain, the American River would've overtopped the levees," Johnson says.

So what (nearly) went wrong?

The first problem is Sacramento was built on a flood plain. The Sacramento and American Rivers collect rain and snowmelt from almost 30,000 square miles of Northern California (what we call the rivers' *watershed*).

All of that water passes right through Sacramento on its way to the Delta and the Pacific Ocean.

We've constrained our rivers into fixed channels using levees, cutting them off from their naturally expansive flood plains. The American River levees squeeze the water into a channel much narrower than the river's natural width. The reason? A hundred and fifty years ago, hydraulic gold mining in the foothills swept down massive amounts of rocky debris that threatened to clog the river. By pinching the American into a narrow passageway, the water flowed faster and harder, sweeping away debris and keeping the channel clear. This worked but created additional flooding risks.

Sacramento's last levees were completed in 1958 and the city thought it was sitting pretty with "300-year flood" protection, that is, protection against a flood so big that statistically it was likely to happen only once every 300 years.

Two things disrupted this cozy picture: Hurricane Katrina, and a changing climate.

The flooding after Katrina revealed that Sacramento faces a potent, previously underappreciated risk: levee failure due to *undeseepage*. The more obvious type of levee failure is *overtopping*, when a river rises higher than the levee and spills over the top. Underseepage is a sneak attack from below, via buried riverbeds left from old river channels. When the river is at flood stage, pressure forces water *under* the levee, through these lost sand and gravel riverbeds. Generally this won't be a huge amount of water, but the flow carries with it some of the sand from underground. Where the sand used to be, supporting the weight of the levee above, a gap is created. At some point, the levee will crack or sink into the gap, and the river will breach it. This happened in New Orleans.

Taking into account the risk of underseepage, Sacramento's overall rating dropped to 100-year flood protection *at best*. According to Johnson, that makes us "the most at-risk urban area in the country for riverine flooding."

SAFCA took action. Through a complex web of financial, legislative, and administrative arrangements, levee improvements began in the late 1990s. To prevent underseepage, a "slurry wall" of impermeable bentonite clay is built into a levee. This wall cuts off the underground flow of river water that might otherwise undermine it. Building such a wall is a massive project, as depending on the geology of a particular site, the wall might have to go as deep as 150 feet. A trench is dug in the levee and filled with a slurry that solidifies to form the wall. A $300 million project to do this on the banks of the American River through Arden/Carmichael (including levees around Jacob Lane and Howe Avenue) was finished in 2016.

Overall about $4 billion will be spent to bring the Sacramento area to 250-year flood protection, with the goal of eventually reaching the 400- to 500-year level.

Improving the levees is necessary but not sufficient. We've learned that peak flows in our local rivers can get a lot bigger than people thought when engineers designed the region's flood control system in the 1950s. Managing those flows in the American River, and adjusting to the changing climate, is the job of Folsom Dam.

Science in the Neighborhood: Weather

Folsom Dam

Levees, those earthen walls along the American River, do a fine job of keeping the residential neighborhoods of Arden-Arcade and Carmichael dry as long as the amount of water flowing in the American doesn't exceed 115,000 cubic feet per second (cfs).

But more than a dozen times since 1905, flows from 20% to 150% greater than that have poured into the American River from its two-thousand-square-mile watershed. What protects Sacramento at such times? Folsom Dam.

According to Rick Johnson, Executive Director of the Sacramento Area Flood Control Agency (SAFCA), Folsom is "the backbone of our flood control system." Folsom is actually a complex of eight dikes and four dam structures located immediately downstream from the confluence of the North and South Forks of the American River. Folsom Dam has nine different official purposes, including municipal water storage, hydroelectric power generation, irrigation, recreation, and fishery management.

Flood control is the dam's only job with immediate life-or-death consequences.

Before the dam was built in the 1950s, the American River's flow would vary tremendously. During heavy rain or snowmelt, vast quantities of water could enter the river in a short time and flood downstream. Folsom Dam blocks those surges and holds the excess water in the reservoir of Folsom Lake. This provides the double benefit of protecting people and property downriver, and storing water for the dry season.

Unfortunately, water storage and flood protection are

contradictory goals. For flood control, the perfect dam would have an empty reservoir behind it, leaving plenty of room to hold a winter deluge. For water storage, the perfect dam would have a full reservoir, maximizing the amount of water available for the summer.

Day by day (or hour by hour in times of crisis) people decide how much water to keep at Folsom, and how much to release. Keep too little, and farmers, fish, and municipalities suffer when there's not enough water. Keep too much, and a big storm could overwhelm the dam.

The dam's managers are keenly aware of how that could happen, because in February 1986, it nearly did.

The floods of 1986 were caused by back-to-back *pineapple express* storms. Also called *atmospheric rivers*, these tropical storms carry warm, moisture-laden air from the Pacific and dump it, firehose-style, over Northern California. For the ten days of the storm period in '86, sites in the watersheds of the American and Sacramento Rivers recorded 30, 40, even 50 inches of rain. Adding to this unprecedented precipitation, the warm air of these storms brought snow only to high elevations (above 7000 feet), and actually *melted* existing snow lower down.

Water flow into the American River peaked at over 250,000 cfs. When Folsom Dam was built, the highest peak flow on record was only about half that. Sacramento's dam and levees were not designed to handle a weather event of this magnitude. Folsom Lake filled, and kept rising.

Dam officials faced an impossible choice: release water into the river and risk overwhelming the levees downstream, or risk overtopping the dam.

Science in the Neighborhood: Weather

The decision was made to release, and the American River came within *six inches* of overtopping the levees.

Extreme weather events like the storms of '86 have become more common in the past 50 years. Scientists using tree rings to study ancient weather patterns now say that California weather in the first half of the 20th century, upon which Sacramento's flood control system was based, was aberrantly mild. Folsom Dam, "backbone" of the system, isn't steely enough.

One proposed solution to Folsom's shortcomings was to construct an additional dam on the American, at the confluence of the North and Middle Forks. This controversial Auburn dam has a long, contentious local history and looks unlikely to be built. (The famously high Foresthill Bridge was designed to accommodate the planned dam and reservoir.) So the existing dam complex at Folsom was modified in a project completed in 2017. The main channel dam and other structures were raised 3.5 feet, and a whole new concrete dam and spillway were built.

The new dam provides a way to increase flood control storage space (that is, more room for floodwater) in anticipation of a major storm. In the original Folsom Dam, water can be released through eight small outlets that only discharge up to 27,000 cfs. Bigger flows are only possible through large gates located near the top of the dam. This means that managers can only dump a lot of water fast when the reservoir is already mostly full. The new dam has big gates at the *bottom*, so large releases (up to the levees' capacity of 115,000 cfs) can begin *before* the reservoir is dangerously high. In a race against rising water, this gives dam managers a head start.

During the peak flood season (November to March), Folsom is kept 40-60% empty. Accurate predictions of future water flows are crucial to effectively managing the reservoir. How do flood managers know when to brace themselves for a major event? Let's find out.

Forecasting Floods

Early Sacramento grew in a floodplain at the confluence of two major rivers because of gold, not because the location was a wise choice. Floods or the threat of floods have plagued the city ever since its founding.

Folsom Dam can stop the American River from flooding Sacramento only if there's enough room in Folsom Lake to hold the water raging down from the mountains. During an extreme weather event, peak flow into Folsom can be more than triple the amount that can be safely released into the lower river. To make sure the water doesn't overtop the dam, it may be necessary to draw down the reservoir in advance. But not too much—water released now is water the community won't have later, in summer.

Therefore, good flood control requires good forecasting.

The Bureau of Reclamation, which manages Folsom Dam and reservoir, has a Joint Operations Center in Sacramento where the National Weather Service, the California-Nevada River Forecast Center (CNRFC), and the California Department of Water Resources (DWR) Flood Operations Center all share one roof. These federal and state agencies work together to collect hydrologic data, build models, and make predictions of how much water, where, and when.

Science in the Neighborhood: Weather

Weather forecasting is the first step. At the Joint Operations Center (JOC), meteorologists watch for storms, especially the big "pineapple express" systems. Five-day precipitation forecasts are pretty accurate and allow enough time to make room in the reservoir if necessary. Actual precipitation is measured by a network of sensors placed throughout the watersheds of the American and Sacramento Rivers. Most of the sensors transmit their data by satellite directly to the JOC.

Measuring the amount of rainfall doesn't directly tell you how much the river is going to rise in response. According to Boone Lek, a senior water resources engineer in the Division of Flood Management at DWR, rainfall data from many locations are fed into mathematical models calibrated to predict timing and magnitude of runoff for the watershed. In some of the smaller watersheds in our area, such as the Cosumnes and Napa Rivers, runoff is fast and floods can be sudden after rain. In the American River, runoff from the peaks of the Sierra to Folsom takes less than a day.

Actual, not predicted, flows are measured by stream gauges maintained by the US Geological Survey. Gauges measure the depth of the water at cross section points along the river or stream and take an average. For each gauge, such as the corrugated metal tower you may have seen just upstream of the I Street bridge, scientists have prepared rating tables that translate water depth at that specific location to flow in cubic feet per second, such as, 10 ft = 50,000 cfs. The cfs measure then gives planners a flood stage rating.

Yet more types of hydrologic data are collected to improve flood forecasts. If you've been hiking in wilderness areas of the Sierra or the Trinity Alps, you might have stumbled on an odd metal hut, A-frame in shape, ten or more feet tall, possibly with adjacent towers and antennas. Here, government agencies and also utilities like SMUD and PGE gather data on temperature, wind, snow depth, snow water content, and soil moisture content. All of these factors affect how much water will runoff into the rivers. The CNRFC posts a ton of this information in visually appealing map form on their website. If you're a weather-watching junkie, you'll love www.cnrfc.noaa.gov, or cdec.water.ca.gov which gives hourly details on reservoir storage across the state.

Even with thousands of sensors, there are gaps in the data network. Citizen scientists can help fill the gaps in official sensor reporting via the Community Collaborative Rain, Hail and Snow Network (www.cocorahs.org). Volunteers submit local precipitation measurements to a national database. To join this group, all you need is a rain gauge and a little training.

So what happens when all the signs indicate a flood threat?

For every reservoir that serves both as water storage and flood control, a *top of conservation level* is calculated. This is the level of water predicted to allow enough space for flood control without wasting stored water. This level is not constant, but is changed depending on time of year, how saturated the watershed is, and more. If a reservoir is higher than its TOC level, then water is released, even in periods of apparent drought. As the climate changes, choosing good TOC levels is getting harder. Rick Johnson of the Sacramento Area Flood Control Agency says, "In the last 25 years, water is coming off the watershed differently." In the past, California's rainy season was generally over before the Sierra snowpack melted. Water managers could keep the reservoirs relatively empty during storms knowing they would fill again with snowmelt later. Nowadays, the Sierra snowpack is melting earlier, *during* flood season. If the reservoirs must be kept partially empty for flood protection, then early snowmelt water is lost and we roll into summer with less water stored. Johnson mentioned another climate change challenge as well: an increase in extreme weather events. "We're seeing more outliers, which the infrastructure was not built to handle."

During a flood the levee-defined main channels of the American and Sacramento Rivers actually carry only a small fraction (15-20%) of the total water flow. The rest is funneled into a system of flood basins called bypasses using mini-dams called *weirs* that are designed to be overtopped at flood stage. The Sacramento Weir is an old (1916) concrete structure with manually operated gates located about four miles upstream of the Tower Bridge. When the I Street gauge reads 27.5 feet and rising, river forecasters open the gates of the weir and send floodwater into the Yolo Bypass (part of which is crossed by I-80's Yolo Causeway bridge) by way of the Sacramento Bypass. (Interestingly, the Sacramento Weir is located *upstream* of the mouth of the American River, because during a major flood event American River water entering the Sacramento backs up and actually flows upstream.)

If homes and lives are in danger, a decision to evacuate will be broadcast to local residents by reverse 911 calls and by sending emergency messages to all cell phones within range of specific cell towers. Rick Johnson warns,

"When they say go, *go*." He advises Sacramentans to pay attention when large storms come. "People in San Francisco know about earthquakes. In the foothills they know about fire. In Sacramento, people have to realize we're living behind levees."

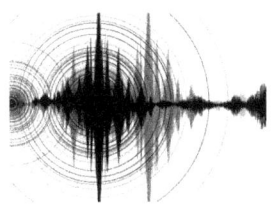

It's not my fault: Earthquakes in Sacramento

My relatives in the Midwest take solace from their winter storms and tornadoes with the knowledge that at least they don't have *earthquakes*. Which is not entirely true: the largest earthquake ever in the lower 48 states shook New Madrid, Missouri in 1812. But there's no denying the ongoing quake risk is higher in California than anywhere else in the US.
What about Sacramento?

In October, millions of Californians participated in the Great California ShakeOut, a day to practice earthquake preparedness. Fewer than 5% of the participants were in the region around Sacramento. Should more of us have paid attention?

According to Chris Nance of the California Earthquake Authority, nowhere in California is free from earthquake risk. However, some places are obviously more at risk than others. Sacramento is in the middle of the most seismically quiet region of the state. Sacramento County does not sit atop any known faults, but we are surrounded by them. To our west where the Central Valley meets the Coast Range, a complex zone of faults lies underground; a strong quake (magnitude 6.6) struck the Vacaville-Winters area in 1892. And of course the San Andreas and Hayward Faults in the Bay Area have the potential to release major earthquakes. To our east, the Foothill Fault Zone runs along the edge of the Sierra Nevada from about Oroville in the north (big shaker in 1975) all the way toward Fresno. The Tahoe region is also surprisingly active, with at least three quakes >6 since Gold Rush times, and there have been loads of quakes around Markleeville.

Nevertheless, Nance says, "We need to acknowledge that we're vulnerable to earthquake damage in this area." Most of us think we can't be affected by a natural disaster. "A lot of people in Houston right now {after Hurricane Harvey} are saying, boy, are you wrong."

If there aren't any faults under our feet, what's to worry about?

The first issue is the possibility that we're wrong. Previously unknown faults trigger earthquakes all the time. And even faults we're aware of can surprise us. The South Napa earthquake of 2014 occurred on a fault that scientists did not think was ready for any shaking.

Second, most of the damage and injury in earthquakes results from seismic shaking. The ground moves, causing things to fall on people, and causing buildings to collapse. How much the ground moves depends on the magnitude of the quake, distance from the earthquake's epicenter, and local geology. The fact is, a major quake miles away can potentially cause significant shaking here in Sacramento. Nance shared this anecdote, which some of you might have experienced too. "In 1989 when the Loma Prieta earthquake struck, I was home at the corner of Robertson and Watt {in Sacramento}. We had white caps in our swimming pool." Depending on how your home was constructed, this type of shaking can cause significant damage. In a worst case, if a house is not reinforced and bolted to the foundation, it will entirely fall off its foundation.

Third, many areas of Sacramento are built on flood plains. The sedimentary soil in these areas is poorly consolidated and sandy. In addition, around the rivers the soil is relatively saturated because the water table (groundwater) is near the surface. These are the two most important risk factors for another type of earthquake damage: liquefaction. Liquefaction is a temporary transformation that occurs when shaking loosens the soil, and groundwater squeezes into the spaces between the grains. Liquefied soil is like quicksand. It loses its ability to support structures built on top, which may sink by inches or more in an irregular way that leads to collapse. Liquefaction can occur at some distance from an epicenter, as evidenced by the damage to San Francisco's Marina District in 1989 after Loma Prieta. Of particular concern in Sacramento, liquefaction could lead to levee failure and flooding.

For these reasons, Sacramento homeowners shouldn't ignore the value of earthquake insurance. The 1994 Northridge earthquake destroyed the commercial market for earthquake insurance, so in 1996 the legislature created the California Earthquake Authority (CEA) as a not-for-profit, publicly managed but privately funded entity to offer earthquake coverage. Policies are sold and serviced by most insurance companies, not by CEA itself. The same coverage is available regardless of who your primary insurer is. CEA has almost $16 billion in claims paying capacity—enough to cover another Northridge or Loma Prieta. Because the known risks in Sacramento are low, so are our rates.

What if you find yourself in an earthquake, either here or while traveling elsewhere in the state? What should you do?

Remember the words "Drop, cover, hold on." Wherever you are, drop to the ground so you don't get thrown off your feet. Then crawl under something sturdy like a table. Hold the table with one arm and put your other arm over your head. Only when the shaking stops should you try to flee a building. As Nance says, "It's stuff falling that hurts people." People who run outside while the ground is still shaking can be injured by debris dropping from the sky.

Whether in anticipation of an earthquake, a flood, or other disaster, everyone should have some kind of family disaster plan and supplies on hand. For help with yours, visit Ready.gov.

Part Three

Utilities

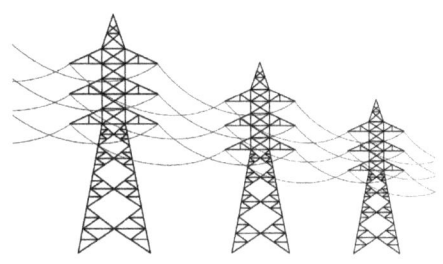

Wrangling Electrons

I'm a member of a local group called Capital Science Communicators. When CapSciComm organized a tour of the Independent System Operator in Folsom, I had never heard of ISO and had no idea what it does. When I had to submit my personal information for security clearance, I knew this was going to be interesting (and possibly a scene in one of my thriller novels). I wasn't disappointed. What the ISO does behind the curtain blew my mind, and is helping to "green" our energy grid.

Of all the complex engineering systems we take for granted in our daily lives, one stands out for our minute-to-minute dependency on it. Unless it's the wee hours of the night, we notice a failure of this system *instantly*.

I'm talking about electricity, of course.

Staggering complexity lies behind the simple act of flipping a light switch. Some of that complexity is managed every second of every day in Folsom at the California Independent System Operator, or ISO. The ISO is an independent non-profit organization, not a government entity, though it is overseen by the Federal Energy Regulatory Commission. ISO manages the flow of electricity across most of California through the high-voltage transmission network—what some people call "the grid."

This is not your neighborhood utility. This is the big stuff, 500-kilovolt power lines that make your hair stand on end if you get close. 26,000 circuit miles of those power lines crisscross the state, connecting over 1900 generation units in California to substations where local utilities handle electricity distribution to consumers.

ISO does not own or physically maintain any of that infrastructure. In the words of Steven Dale Greenlee, an ISO public information officer, ISO's job is like "air traffic control for electrons." ISO controls how much electricity enters the grid, from which power plants, and the route by which it flows to consumers.

This is a lot harder than it sounds. The supply of electricity in the grid must be constantly matched with the demand. At any instant, the amount of electricity being generated and put into the grid (the "load") must equal the amount of electricity being pulled out by consumers ("demand"). If the load is too great, a power surge damages equipment and leads to power outages. Conversely, if the load is insufficient to meet demand, not all customers will get the power they need and there will be brown- or blackouts.

To keep the balance—and the lights on for everyone—ISO adjusts the load *every five seconds*. How?

Science in the Neighborhood: Utilities

A team of ten operators, mostly electrical engineers, work twelve-hour shifts monitoring a massive amount of data coming to the ISO about the status of the grid. Visitors are not allowed on the control room floor, where a dazzling, curved wall of video screens 80 feet long and subdivided into a hundred panels gives form and color to this data. ISO designed a custom "geospatial data presentation" scheme to make it easy for the engineers on duty to see exactly what's happening on the grid at a glance. Interweaving red and blue lines like a circuit diagram from a SciFi movie illustrate the status of substations and circuit breakers and transmission paths across the state.

The engineers' job is to solve a complex puzzle. At any given moment, consumers are using a certain amount of electricity. That electricity comes from a range of generators, mostly powered by natural gas and solar, located all over the West. The electrons have to travel from generator to consumer over the grid. But just like a water pipe, the grid's transmission lines have a maximum capacity. ISO figures out how to route electricity so no particular line gets overloaded. They're also ready to instantaneously work around local failures in the system, always planning with two layers of contingency. If the original route *and* the re-route fail, a second alternative way to get the power to where it's needed is programmed in.

But redirecting the path electricity is flowing won't help if there isn't enough juice to go around. So as part of the agreement ISO has with power generators who feed the grid, power sources must be "dispatchable." That means ISO gets continuous, automatic control of the amount of electricity being produced. To match demand,

ISO can turn up or turn down the burning of natural gas at a power plant hundreds of miles away.

This can't be done instantaneously. It takes time to start and stop older, conventional generators, so another part of ISO's job is to *predict* energy demand. Seasonal patterns, weather forecasts, requests from the utilities and more are used to optimize the load 24 hours in advance. Power plants come online early, in anticipation of demand.

If the demand forecast is wrong—for example, the weather turns out hotter than predicted and air conditioning units are running long and hard—ISO has to figure out where to get the extra power, and how to carry it to consumers across transmission lines that may already be at capacity. In a pinch, ISO calls a Flex Alert. This is a voluntary request for people to conserve electricity and relieve some of the pressure on the grid.

In parallel with regulating the generation and flow of electricity, Cal-ISO manages an electricity pricing market that puts a fluctuating price on every kilowatt running through the grid. The price changes every five minutes, and is an important tool for ISO to encourage conservation or consumption to match demand with supply. Large commercial users can adjust their power usage in order to save money. (In the early days of this energy market, the infamous company Enron gamed the system, driving up prices and contributing to the California energy crisis of 2000-2001. Since then, the market has been tightly regulated to prevent such abuses.)

Renewable power sources, especially solar and wind, are contributing more and more of our state's electricity. Solar and wind power offer advantages. They don't contribute to carbon dioxide emissions, and they're virtually free once the infrastructure is in place because you don't have to buy fuel. But converting the state's power grid to rely on renewables presents significant technical challenges for the ISO.

The number one problem with renewables is they're *variable* power sources that we can't turn on and off. Only Mother Nature gets to decide when the wind blows or the sun shines. At 3 PM on a July day a large fraction of our electricity needs could be met by solar power alone, but a few hours later after the sun sets, that power generation drops to zero. And unlike fossil fuels, electricity cannot be stored for later use. Solar power has to be used when it's produced. You can't stock up and use it after dark; you must have access to another generator.

Why not? Why not use batteries to store electricity made while the sun is shining? That would be ideal, but we can't—not yet, anyway. While batteries are fine to power your mobile phone, the scale of electricity storage needed for the grid is way beyond current battery technology. Energy companies are investigating ways to store power without using chemical batteries—innovations like compressed air as a source of potential energy. (You compress the air using solar electricity, then you later use the compressed air to drive electricity generation.) Solving the storage problem would pave the way for renewables to really take over the grid.

One alternative to storage is to shift time of consumption to match generation. Peak solar power production is highest between noon and 4 PM; wind tends to be strongest at night and early morning. Unhelpfully, peak power usage is between 5 and 8 PM. So one way to integrate more renewables into the grid is to shift consumption away from those peak evening hours. Consumers can help by running major appliances around noon and never in the evening, or cooling their homes a bit extra in the afternoon so their air conditioners don't have to run as hard after 5 PM. Anything that reduces the peak load in the evening takes pressure off the grid, and means fewer conventional power plants are needed.

Sacramento is a leader in the greening of the grid. If you live in the city or county of Sacramento, your electricity is supplied by SMUD (Sacramento Metropolitan Utility District). SMUD is a public, municipal utility structured as a not-for-profit, and is governed by a board of seven directors. Thanks to this arrangement, SMUD customers pay about 30% less for electricity than our regional neighbors who are served by investor-owned Pacific Gas & Electric.

Reflecting the priorities of many local residents, SMUD's board has a history of emphasizing environmental goals even before government regulations make such goals mandatory. SMUD was the first utility in California to reach 20% of power produced from clean, renewable sources—and that percentage keeps rising. Today, 40-50% of SMUD electricity is generated from non-carbon emitting sources: hydroelectric, solar, wind, and biofuel. (The remainder comes almost entirely from natural gas-fired power plants. There is *no* coal-fired generation in the state of California.)

This puts SMUD somewhat ahead of the curve as the state toughens laws regulating how utilities may source their electricity. The California Renewable Portfolio Standard currently mandates that 33% of the state's power must come from solar, wind, biofuel, and "small" hydroelectric by 2020, and 50% by 2030. (For some reason clean power from "large" hydroelectric plants is a separate category, so actual renewable generation will be even greater.) In 2017, a sunny, windy year with full water reservoirs, California reached about 32% renewable electricity, and for one spectacular day—May 13—a recordbreaking 67% of California's power came from renewables.

SMUD is growing its solar generating capacity. In a 1989 referendum, SMUD ratepayers voted to shutter the utility's only nuclear power plant, at Rancho Seco. SMUD then built a natural gas plant at the site to take advantage of existing power lines and infrastructure. But there was still plenty of room for a large solar farm. SMUD completed the Rancho Seco Solar Array in 2016. Solar panels built on rotating turrets follow the sun, generating up to 10.5 MW of clean, renewable power. Some of that power is earmarked for the Golden 1 Arena, which is 100% powered by the sun. 85% of the arena's electricity comes from the Rancho Seco solar project. The rest is generated from rooftop panels onsite.

Consumer demand is driving SMUD's investment in renewable power generation. Golden 1 isn't SMUD's only big customer signing up for green electricity. UC Davis Medical Center signed an agreement to buy a percentage of their power as solar in order to reduce the center's carbon emissions. Individual consumers can also green their electricity consumption without having to build a solar system on their home by buying into SMUD's renewable power program called Greenergy. For only a few dollars extra per month, you can have 50% or 100%

of your power generated from renewable resources.

SMUD is rolling out a plan to use price incentives to shift toward renewables, and to reflect the actual cost of electricity at different times of day. Starting in late 2018-2019, SMUD will transition customers to time-of-day pricing. The rate you pay per kilowatt hour will vary depending on when you use it, with the highest price during the evening peak. (SMUD's current "tier" pricing, which is based on how much power you use regardless of what time of day you use it, will end.) If power consumption can be spread more evenly through the day, this will support our state's shift from natural gas to renewable electricity.

From river to tap: Water supply

> Water issues in California are notoriously complex and contentious—at least, the legal aspects of how to divvy up our limited supply. I stuck with some simpler local questions: Where does my water come from? Is it clean? And why do I still have water during a power outage?

Ground water

High-quality drinking water, flowing from the tap, is a luxury we often take for granted in the US. According to the World Health Organization, at least 2 billion people worldwide don't drink clean water. So where does our tap water come from, and how is it kept clean and safe?

There are many different answers to that question in

the Sacramento region. Unlike wastewater treatment, which was centralized under Regional San in 1973, clean water is delivered to Sacramento County homes by a patchwork of 24 local water districts. These districts originated from early property developers drilling wells, and expanded naturally into public water agencies. Some have consolidated over time. For example, Sacramento Suburban Water District (SSWD), one of the largest local providers that serves much of Arden-Arcade and areas north of I-80, was formed by the union of Arcade and Northridge districts in 2002. Mergers must be approved by elected officials and are not always successful; an effort to unite SSWD and Placer County's San Juan Water District failed a few years ago.

Each water district has its own legal structure, manages its own system of pipes, and has its own water sources. Water sources can be either surface water (water from local rivers and reservoirs) or groundwater (water pumped up from underground basins). For example, some districts rely 100% on surface water from Folsom Lake, which puts them at risk during severe drought. Arden-Arcade's SSWD has access to both surface and groundwater. SSWD operates 75 active wells and also has contract rights with the City of Sacramento and the Placer County Water Agency to receive surface water from the American River and Folsom Lake.

How does SSWD decide whether to provide groundwater or surface water to its customers? According to General Manager Dan York, groundwater is cheaper but whenever possible, SSWD turns off its groundwater wells and delivers surface water. Why? Groundwater is a limited resource that gets depleted if overused. In fact, from the 1950s the groundwater table in SSWD's service area was dropping at the astonishing rate of about one and a half feet per year. Switching to surface water during wet times "banks" groundwater and even allows the underground basin to recharge. Because of this policy York says they've actually seen some recovery in the depth of the local groundwater table, which averages about 150 feet in Arden-Arcade.

SSWD's wells are located on small lots owned by the district. York took me to visit a typical example. These nondescript, fenced sites are located all over the area. I was surprised to discover near my home several wells I'd seen but never recognized, next to commercial buildings, or on property behind a residence. Wells are drilled from 300 to 1000 feet deep. To comply with state law, SSWD monitors water quality by collecting over 2,500 samples per year for chemical testing, as well as thirty bacteriological samples every week from the distribution system. Soil contains many naturally occurring contaminants, so sometimes water from a well will show an excess of a regulated level of a contaminant. Because each well is independent of the others, SSWD normally will turn off the contaminated well and rely on other sources until the water is treated or the contaminant is no longer detected.

Private property owners are allowed, with a permit from the County, to drill their own personal wells. They're not required to do any water testing, so York thinks it would be unwise to drink such water, though it could be used for irrigation. Backflow devices prevent customer-drilled water from entering the water pipe system, which contains only water that has been tested and is potable.

SSWD's ground water is raised to the surface by vertical turbine pumps with capacities from 300 gal/min to 3500 gal/min. The pumps are powered by electricity. I asked York why my home has water even during a power outage. Apparently that wasn't always the case. During California blackouts in 1996, Sacramento residents lost water along with power. Because of this hard lesson, SSWD invested in emergency diesel- and natural gas-powered backup generators throughout the district. With power from these generators, they can pump enough water to provide for health and safety even during a widespread power outage. (Not all water districts have this capacity.)

From their headquarters on Marconi Avenue, SSWD can remotely monitor all their wells, and adjust settings without having to go to the well. The main issue is water pressure. SSWD tries to maintain a water pressure of 30-60 psi in the system. At each active well, if the psi drops to 30, the pump turns on. At 60 psi, the pump shuts off.

As water is pumped from each well, it passes through a sand separator using gravity to remove any grit. Then SSWD adds chlorine as a disinfectant. In part of the district (their south service area), fluoride is also added. Other than that, the groundwater delivered to your home is virtually unchanged from the water that came out of the earth. This makes groundwater cheaper for the district to supply than surface water—water from the American and Sacramento Rivers requires more costly treatment to make it safe to drink. How is that done? Read on!

Science in the Neighborhood: Utilities

Surface water

In Sacramento we live on the banks of two beautiful rivers. Maybe you've gone swimming in one of them, or boating, or fishing. But have you ever drunk the water?

Don't be too quick to respond. The answer is almost certainly yes.

The American and Sacramento Rivers provide about 80% of the tap water to the City of Sacramento. (Residents of the county also consume surface water from the rivers, but the exact mix of groundwater versus surface water varies depending on where you live.) Of course the water is not piped directly from the river to your home. Despite a current fad for "raw water" among a fringe of believers, drinking straight from the river would put you at risk of waterborne infectious diseases. River water is treated first to make it clean, clear, and safe.

The City of Sacramento Department of Utilities runs two surface water treatment plants. The Sacramento River plant, located near Richards Boulevard, began operating back in 1924 and was entirely renovated as of 2017 to treat 160 million gallons of water in a day. The American River water treatment plant, named E.A. Fairbairn, went live in 1964. Both conventional water treatment facilities are notable for a large and stylish intake building that's clearly visible on the rivers (near the old Powerhouse, and near Sac State). Both facilities are connected to a series of 1600 miles of water lines in the City that serve nearly half a million residents. In winter, when water demand is low, either plant alone can meet the city's needs, so a plant can be shut down for maintenance.

Brett Ewart, a senior engineer for the City, gave me

a tour to learn how the Fairbairn plant on the American River works. As I admired a kingfisher along the riverbank, Ewart explained why the intake structure is so big. "The intake is designed to house a series of eight pumps that lift the water up from the river. Flow through the treatment plant then works by gravity," he said. "At maximum capacity, we could pull 200 million gallons of water a day through the intake, though the most we've ever actually treated and sent out is closer to 100 million gallons {over one day}." If all that water entered through a single intake of modest size, the strong sucking pull would be dangerous. Instead, the intake was designed with a large surface area for the river water to flow into. This keeps the velocity of the water low (less than 0.33 feet per second), and protects the river's fish—in particular, the American River's fragile populations of migratory salmon and steelhead. The intake pumps themselves are covered by fish screens fine enough to exclude juvenile fish and allow them to swim away on their journey to the ocean. To further protect the river ecosystem, at times when the fish life cycles are particularly sensitive, the City limits the amount of water that can be diverted from the river. Any shortfall is made up with groundwater or water from the other plant.

Raw river water flows downhill to the first phase of the treatment process. In a giant settling tank or "grit basin," the water is allowed to stand rather still to allow larger particles of debris and sand to drop out and settle on the bottom. Next, the water flows through inlet meter pipes that measure how much water has been extracted from the river. Samples are taken continuously to monitor the ever-changing turbidity and composition of the river.

Most samples are analyzed at the on-site laboratory.

At this point the water is still full of suspended particulates. In the next treatment phase, those tiny particles are encouraged to clump and fall out of solution by adding *coagulants* to the water. The silt and clay that make up a lot of the suspended material generally carry a negative charge, so the particles repel each other. By adding a positively charged coagulant such as alum (aluminum sulfate), the charge repulsion is overcome and the particles can clump together. This happens with slow, gentle mixing as the water passes through a serpentine path of *flocculation tanks*.

Once the suspended material in the river water has flocculated into heavier clusters, the clusters (and the added coagulants) are allowed to settle out of the water in a wide, calm *sedimentation basin*. I could easily see how the water progressed from cloudy to clear as I walked from one end to the other of a pathway above this basin.

Still moving with the flow of gravity, the visibly clear water now enters the filtration stage. Ewart points out that with groundwater, this process happens naturally. At the plant, water sinks down through a bed of anthracite and sand layers in a gradation of sizes that get finer as you descend. Now the turbidity is gone and many microbial contaminants have also been removed because they cling to particles in the water. To kill any lingering hazardous microorganisms (bacteria, viruses), the water is chlorinated for a period of time. Finally, fluoride is added for dental health in the community, and the water's pH is balanced with lime in order to protect pipes and water delivery infrastructure from too much acidity.

A final set of pumps pressurizes the water to drive it

directly to household taps, or to eleven giant freshwater storage tanks sprinkled throughout the City above or below ground. Each tank holds 3 to 15 million gallons of water. One of these you've seen while driving on the Cap City Freeway. Called the Alhambra Reservoir, it's decorated by 2000 feet of blue LED lights. The illuminated artwork represents the grid of city streets on one side, and the river flowing in from the other. Next time, pay attention to the blue circle in between: the lights in the circle rise and fall, representing the water level in the tank. Two additional tanks of the same design are found near Sac City College and the UC Davis Medical Center.

Sacramento's drinking water meets or exceeds government standards. Wondering what that means, exactly? Consumers get an annual water quality report on this data sent to them every June. You can view the latest Consumer Confidence Report online at cityofsacramento.org/Utilities/Resources/Reports.

After the Flush:
Where does your wastewater go?

A few years ago my family and I rafted the Colorado River through the Grand Canyon. For two weeks we were cut off from civilization. We had to bring with us everything we needed, and we had to carry out everything that we brought. *Everything.* Including, to put it delicately, our solid waste. It actually wasn't a big deal—you can ask me the details—but it got me thinking about sewage management and rivers.

In September 2006, garbage collectors in Sacramento went on strike. For several weeks, homeowners' trash piled up in driveways, yards, and streets. It was a stinky, ugly inconvenience.

Losing solid waste pickup was nothing compared to what would happen if we lost our wastewater disposal system. Can you imagine two weeks without a flush?

Wastewater treatment gets my vote for the most

underappreciated science-based public utility. What it lacks in glamour, it makes up in importance. The British Medical Journal named sanitation the greatest medical advance since 1840 for its role in reducing waterborne diseases. Without good sewage collection and treatment, a community's drinking water cannot be safe.

Prior to the federal Clean Water Act of 1972, the safety of Sacramento's water supply was at risk. The region had 22 small wastewater treatment plants discharging directly into the Sacramento and American Rivers. To protect local waterways and comply with the Clean Water Act, local leaders decided to "regionalize" wastewater treatment. Sacramento County, along with the cities of Sacramento and Folsom, joined together to form the Sacramento Regional County Sanitation District (now known as Regional San). Between 1976 and 1982, Regional San built a single massive wastewater treatment plant in Elk Grove that replaced the older plants and now serves about 1.4 million residents from Folsom to West Sacramento, Citrus Heights to Elk Grove.

The Sacramento Regional Wastewater Treatment Plant operates 24 hours per day, 365 days per year, processing an average of 150 million gallons of sewage per day. Like most Sacramentans, I didn't even know where this 3000+ acre operation was located until [I signed up for a public tour](). Yes, a surprising number of ordinary people want to walk around a wastewater treatment plant, and Regional San obliges with free, seasonal monthly tours.

The plant's site in Elk Grove was carefully chosen. First, it sits at a very low elevation (32 feet below sea level), which saves money because wastewater can flow to the plant by gravity (although some pumping is required in the conveyance system). Second, it is only about two miles from the Sacramento River, where the treated water leaving the plant (called *effluent*) is discharged.

Whatever you flush or send down your sink or shower enters a colossal underground system of pipes. Pipes with a diameter of three or four inches typically carry your wastewater from your home to a grid of larger main lines and trunk lines under the streets. That system of sewer pipelines is managed by your local collection district (such as City of Sacramento or Sacramento Area Sewer District). Local sewer collection pipes funnel into much larger "interceptor" pipelines which are managed by Regional San, analogous to the way local roads feed into interstate highways. Your utility bill reflects separate charges for the services of your local collection agency and Regional San. Regional San has about 177 miles of pipes compared to the local systems' six or seven thousand miles, but Regional San's interceptors are as large as twelve feet in diameter.

The journey of wastewater from your home to the treatment plant takes time. Morning shower water from Citrus Heights arrives maybe ten hours later; wastewater from Folsom may take a day. (Because of this, flows coming into the plant are generally lowest in the morning.) Along the route, Regional San controls odors at pump stations using chemical scrubbers. The plant itself was designed with hundreds of acres of open space around it. These "Bufferlands" protect Elk Grove residents from the stench. (I was lucky to visit on a windy day—the plant hardly smelled at all.)

On the tour I saw a jar of *influent* (wastewater that

enters the plant). It didn't look like my idea of sewage. Influent is watery, cloudy, slightly yellow-gray in color, and a bit sandy. It can, however, carry debris such as rocks and tree limbs that would damage the plant's equipment. Therefore the first step in processing wastewater is to remove bulky debris and send it to a landfill. This isn't a delicate operation. Screens made of metal bars ten feet long and as thick as your finger act as filters.

After bulky debris is removed, the wastewater enters the influent pumps. Sitting at the lowest point in the plant, these massive pumps can lift as much as 125 million gallons per day, thirty-five feet up. Gravity then pulls the wastewater through treatment and release.

Next comes the actual wastewater treatment. How do they clean wastewater to meet legal standards for discharge into the river? With clever engineering, some chemistry, and help from a lot of very little friends.

Treating wastewater

Every time you flush, that stuff has to go somewhere. Wastewater from 1.4 million people in the Sacramento region ends up in the Sacramento River by way of the Sacramento Regional County Sanitation District (Regional San) wastewater treatment plant in Elk Grove. At the plant, people like engineer Ruben Robles oversee the transformation of liquid yuck into water that by some measures is cleaner than the river into which it's discharged.

Robles is Director of Operations at Regional San. He emphasized to me how fortunate we are to live in a place with the money and organization to protect our health and environment from raw sewage. "We do an exceptional job with wastewater treatment in the US," Robles says. He ought to know. He's visited wastewater treatment facilities around the world, even opting for the Paris sewer tour on vacation while his wife visited the Louvre. (That's dedication!) Robles guided me through the processes at the plant, saying, "Currently we perform primary and secondary treatments, and also do some water recycling."

First, heavier material like sand and grit is physically separated from the wastewater. Then the wastewater enters primary treatment where solids and lighter materials (like oils and fats) are removed. During this step, wastewater is pumped into very large tanks where it moves slowly. Fats float to the surface and are skimmed off. The solids sink, and are collected and removed at the bottom. The remaining cleaner water moves on to secondary treatment.

Secondary treatment is a biological process that harnesses the extraordinary metabolic powers of microorganisms. In this phase, wastewater is mixed with an activated sludge of bacteria, protozoa, and tiny animals called rotifers. In a microscopic example of "one man's trash is another man's treasure," these diverse microbes feast on organic molecules in the sewage, breaking them down. The microbes and organic matter settle out and are removed, cleaning the water.

While microbes are tackling the organic waste, humans at the plant are working to keep the microbes happy. That primarily means keeping them flooded with oxygen, as the biochemical reactions they perform are *aerobic*. Regional San produces pure oxygen on site by *fractionation* of air. Air is compressed and cooled until it turns into a liquid. The different components of air have different boiling points, so as the liquid air warms up, oxygen can be separated from the other gases. Pure oxygen is then bubbled through the wastewater with the activated sludge to feed the microbes.

When the system is healthy, activated sludge microorganisms form interdependent food chains and food webs. Scientists in Regional San's on-site laboratory keep a close eye on the activated sludge population, measuring the number and types of organisms every day. Sometimes the ecosystem in a treatment plant goes out of whack and the wrong kinds of bacteria take over. During my tour of the lab, Dr. Srivi Ramamoorthy, the laboratory manager, said fixing it is "like baking bread." I laughed. How in the world is sewage treatment like baking bread? "You know what sourdough starter is?" she asked. "We use something similar." Extracts of old, healthy sludge can restore the desired microbial community to the tanks at a plant. Wastewater treatment plants even exchange this "starter kit" with one another if needed.

After secondary treatment by microbes, chlorine is added to the wastewater to kill disease-causing bacteria and viruses. Then the chlorine is neutralized and the now-clean water is discharged into the Sacramento River via a ten-foot diameter diffuser pipe that lies across the bottom of the river. The entire process at the plant only takes about eight hours.

Solids removed during primary processing do not go to the river. About 75 tons of solids are produced every day at the plant. After being separated from the wastewater, solids are sent to huge (45 feet deep) enclosed biological reactor tanks where once again, microorganisms are responsible for the chemical breakdown of the waste. The kinds of bacteria involved and the reactions they perform are different from the activated sludge used during secondary treatment of the water. For solids, *anaerobic* digestion is the rule. In the absence of oxygen, bacteria break down the organic molecules in the waste. One byproduct of this anaerobic digestion is methane gas. This valuable component of natural gas is collected and burned to produce electricity at an on site cogeneration plant operated by SMUD (similar to what happens with biogas at Kiefer Landfill).

After about fifteen days in the anaerobic digesters, the sludge has been transformed into "biosolids" which are pumped to storage basins, basically giant ponds covered with a layer of water for odor control, where the solids will remain for as long as five years. At that time the biosolids are high in nitrogen and also contain some salts and metals. About three-quarters of this material is permanently disposed of on site on land lined to protect the water table below. The remaining one-quarter is recycled into pellets for use primarily as an agricultural fertilizer. Federal regulations set standards for the proper application of this fertilizer to prevent the buildup of too much salt and metal in the soil.

Wastewater processing in Sacramento is already very good but it's about to get even better. To comply with new state requirements, Regional San is adding tertiary processing. The massive "EchoWater project" is underway!

The EchoWater Project

Wastewater in the Sacramento region is processed at Regional San's massive treatment plant in Elk Grove. Every day an average of 150 million gallons of raw sewage is processed into an effluent that gets discharged into the Sacramento River. Local residents might ask: is that processed wastewater clean enough?

There are lots of different ways to define "clean," and some definitions are laid out by the government in discharge permits. Regional San's discharge permit is issued by the State of California in accordance with state and federal law (such as the Clean Water Act), and must be renewed every five years. Contrary to what I expected, the standards for effluent are not the same for every wastewater treatment facility. Instead, specific water quality standards depend on the beneficial uses of the water into which the effluent is discharged. The Sacramento River is a treasured local resource. It's used for agriculture, recreation, and as a source of drinking water for other parts of California. Therefore the standards in Regional San's discharge permit are quite high, much higher than at facilities that discharge into, for example, the Pacific Ocean.

To assure that the standards are met, a state-certified testing lab is on site at the treatment plant. The lab employs biologists, chemists and analysts who perform about 60,000 tests per year, looking at everything from pH and phosphates to organic substances. Every day the lab tests the plant's effluent for coliform bacteria, ammonia, and suspended solids. The lab also tests for trace amounts of metals and organics in special positive pressure rooms designed to prevent contamination.

Fluorescent lights are even banned in one room because such lights contain mercury and at the miniscule levels being studied, this can affect the test results.

In addition to testing for specific chemicals in the water, the lab performs bioassays to see if the water affects the survival, growth, and reproduction of river organisms. In one bioassay, tiny rainbow trout (less than one month old) are placed in tanks of effluent from the plant for four days, and survival rates are measured. Regional San's permit requires that at least 70% of the fish be able to survive in effluent water.

By some measures, Regional San's effluent is cleaner than the river water it joins. It's less turbid, and has fewer bacteria. By other measures, the effluent is a pollutant. In particular, treated wastewater carries nitrogen-containing compounds such as ammonia. Ammonia can be directly toxic to fish. It also acts as a fertilizer to promote the growth of algae and bacteria. Overgrowth by these microorganisms depletes oxygen from the water, causing further harm to the ecosystem.

That's about to change. When Regional San's discharge permit came up for renewal in 2010, new, stringent requirements were added to protect the river and the Delta, whose fragile ecosystem receives flows from the Sacramento. By 2021, the plant must add a tertiary treatment process that will reduce its ammonia discharge by nearly 95%. Also, by 2023 they will further reduce disease-causing microorganisms (especially viruses) by adding a step that filters the effluent through sand and anthracite coal. Together these mandatory improvements are called the EchoWater Project.

These improvements don't come cheap. With a budget

of about $2 billion, EchoWater is one of the largest public works projects in the region's history. It's a massive operation that began with construction of a miniature version pilot plant that ran for about two years to test various technologies and systems. Construction of the full-size facility is well underway, currently under budget and on schedule. To pay for EchoWater, Regional San is gradually raising customer rates. Presently $12.50 per month of a single-family residential bill (total $35) goes to fund EchoWater. This amount will increase slightly by 2020. Fortunately these numbers are well below original estimates.

EchoWater will advance our region's treated wastewater but things like phosphorus, pharmaceuticals, and pyrethroids (insecticides) still can pass through the wastewater treatment process. Public education programs encourage people to keep these constituents out of the waste stream in the first place. In case they're required to treat for chemicals like these in the future, Regional San engineers designed EchoWater with room for additional treatment technologies.

In the meantime, one way to keep pharmaceuticals out of our waterways is for all of us to properly dispose of unused drugs at a local collection bin or hazardous waste facility. I asked Ruben Robles, Regional San's Director of Operations, for the one thing he most wanted to tell people. "Don't flush or dump your unused meds down the drain," he said. "They'll end up in the river, and no one wants that."

Garbage:
Out of sight, out of mind

Some weeks, trash pickup day can't come soon enough. It's mind-boggling how much garbage a household produces. But no big deal—I drop bags into my bin, roll it to the street, and my refuse magically disappears. Guess what? It's not magic. In Sacramento the household waste stream is channeled into multiple fates, including landfill, recycling, power generation, and specialized e-waste processing.

Separating trash

Where does your garbage go?

Into the bin, out to the curb, and presto, once a week, it vanishes!

For most of us, our household solid waste is out of sight, out of mind. But Doug Kobold, Program Manager for the county's Department of Waste Management and Recycling, knows the disappearing trick we take for granted relies on good science and engineering—as well as careful financial planning.

Sacramento County's Department of Waste Management and Recycling collects hundreds of thousands of tons per year of virtually every kind of solid waste that area households produce, from used cat litter to grass clippings to batteries. Based on what's safe for the environment and what's economically feasible, different types of waste are handled differently.

Customers of the county's curbside trash pickup program know that the waste "stream" is first separated by people like you and me into containers for green waste (leaves, grass clippings), mixed recycling (which includes clean plastics, paper, glass, and metals), and household garbage. Batteries, medications, paint, motor oil, fluorescent light bulbs (including CFLs, which contain mercury vapor), E-waste, and solvents must not be placed in any of the curbside pickup containers. These and other hazardous wastes can be dropped off (for free) at the North Area Recovery Station near I-80 and Watt. Kobold says, "We rely on individuals to keep toxic materials out of the waste stream. We're trying to spread the word about putting materials in the right place."

By separating our trash, we help the county divert a remarkable 73% of solid waste from going straight to the landfill. Using the recycling bins also helps to keep our bills down. According to Kobold, as the market for recycled materials has matured over the past fifteen years or so, county recycling programs which used to cost them money are now earning income. Contractors buy 36,000 tons of mixed recycling waste from the county every year, at a price from $15 to as much as $45 per ton.

Science in the Neighborhood: Utilities

Components of the mixed recycling bins, or *single stream recycling* waste, do not all have the same value, and cannot be used while still mixed. All of the recycling waste must be sorted, separating the valuable from the worthless, and what's recyclable from what must go to the landfill. This process is partly automated, but actual humans do a lot of the work. People pluck unrecyclable material, or *residual*, from waste rolling by on a conveyor belt. Residual includes items that never should have been put in the recycling bin to begin with, as well as things like paper that's too contaminated or soiled to be recycled.

A variety of automated sorting techniques are also used. *Star screens* are beds of rotating star-shaped disks that separate broken glass from whole bottles or jars by jostling the glass along, shaking out the small bits, which drop through the spaces between the disks. Aluminum cans are sorted by running the waste over an alternating magnetic field that creates an *eddy current* of electricity in the metal and nudges the cans away from the nonmetallic waste.

Metals, especially aluminum, are the most valuable material. It costs much less money and energy to recycle an aluminum can than to make one from mined raw materials. Recycling PET #1 plastic (most soda and water bottles) also saves energy compared to using virgin material. Glass, on the other hand, while infinitely recyclable, is so cheap to produce from its raw material (sand) that there is no profit to the county in recycling it. Nevertheless, recycling keeps glass out of landfills, which has environmental and economic benefits of its own.

Unlike the contents of the mixed recycling bins, green

waste is a liability. Sacramento County must *pay* contractors to get rid of this material. But different components of the green waste have different uses, and some cost more to dispose of than others. Green waste is sorted into big pieces (branches and such) that can be sent to biomass energy plants to be burned as fuel for electricity generation. Finer materials might be put to use as cover at the county's landfill, or shipped to a processor in Durham (near Chico) for composting.

What happens to the rest of our household garbage?

We dig a hole and bury it.

Despite advances in recycling technology and the market for recycled products, much of the stuff we throw away goes to the landfill.

While it sounds simple to dig a hole, fill it with trash, and cover it up, managing a landfill that meets current standards for environmental safety is complex. Next, I'll introduce you to the civil engineer charged with protecting the air, water, and people around Sacramento County's Kiefer Landfill not only today, but for decades into the future.

Landfill Gas Turns Trash into Power

On a dry, sunny day I drove past strawberry farms and fields of California poppies in search of a big, stinking mess.

Kiefer Landfill is located amid rolling meadows and vernal pools east of Sacramento, near Sloughhouse. As the only permitted municipal solid waste landfill in the county, Kiefer Landfill gets virtually all of Sacramento's household garbage, about 600,000 tons of solid waste per year. I thought it would be easy to spot—or smell.

Science in the Neighborhood: Utilities

In fact, if you don't know what you're looking at as you drive in on Kiefer Boulevard, the county-owned landfill site appears as nothing more than a small power plant and a set of truck weighing scales, adjacent to a large, grassy hill. The appearance is clean and natural.

Making sure the underground reality matches this surface appearance is the responsibility of Tim Israel, a senior civil engineer with the county's Department of Waste Management and Recycling. Israel oversees the site's environmental control systems, compliance and monitoring.

I climbed into Israel's mud-splattered pickup for a tour. The truck skirted a massive hill of grass and lupine punctuated by wells sticking like straws out of the earth. Pipes webbed the grass. Hidden underneath lay trash two hundred feet deep.

"The first hole was dug here in 1967," Israel said. "Conservatively, I'd say we have capacity for at least another sixty years. Maybe longer, if we keep diverting more of the waste stream away from the landfill."

The main environmental hazard of a municipal landfill is leaching of chemicals out of the buried trash and into the surrounding soil and ground water. To prevent this, since the early 1990s the County has lined all newly dug pits with a sophisticated geosynthetic lining system. This begins with a layer of very low permeability clay sandwiched between two pieces of resilient fabric called geotextile. Atop this clay liner is a geomembrane, essentially a thick sheet (60 mil) of impermeable HDPE plastic. Additional layers of alternating geomembranes and "geogrids" funnel any water leaching through the garbage into a collection system. This water could carry contamination away from the site and into local water sources, so it is collected and diverted. At Kiefer, they collect 10,000 gallons per day of leachate water into a sump. Then it's pumped back into the waste mass, or into a water truck to spray for dust control.

Only one section of the landfill site is being used as a dump at any time. I knew we were almost there when we passed through a crowd of thousands of seagulls. "They're here in the winter," Israel said. "They leave in May." Atop a gentle slope, bulldozers pushed piles of garbage over the edge. Additional vehicles smoothed and packed the load. Scavenging gulls and blackbirds whirled amid the stench I'd expected to encounter sooner.

"At the end of every day we cover the trash, sometimes as much as two or three acres," Israel said. The daily cover keeps out scavengers, and minimizes odor. Giant reusable tarps on rolls lie next to the trash heap for this purpose. Green waste that's not suitable for composting is used as an alternative cover at times. Every few weeks, a layer of soil is spread over the trash. When an area of the dump has reached its maximum allowed height of 325 feet above sea level, it can be permanently covered by six feet of soil, and planted over with shrubs and perennials.

What happens next is up to the bacteria.

Over time, microbes break down buried waste. How quickly depends on what's in the garbage, the temperature, how much water is present, and other factors that affect microbial growth and metabolism. Decomposition can be very, very slow. Israel tells a story about digging a hole into a five-year-old area of the landfill, which is something they do to bury particularly nasty stuff like illegal marijuana, medical waste, and handguns. There he found a cabbage that looked "like the day it came out of the refrigerator."

When bacteria eat garbage, they produce waste products of their own, just like we produce CO_2 from our food. In a landfill, where little or no oxygen is present (*anaerobic* conditions), many bacteria produce methane—otherwise known as natural gas. That gas has value.

Landfill gas extraction began at Kiefer in 1997 with wells drilled into the waste. The wells must be tightly sealed; if air gets in, the gas is diluted, and it's even possible to start an underground fire. Typically the gas is about 50% methane, mixed with water vapor that must be removed before it can be burned. The landfill gas wells at Kiefer pump enough methane to fuel two on-site power plants with a total generating capacity of 15 megawatts. That's renewable electricity to power about 10,000 homes. SMUD buys the electricity for their Greenergy program. Unlike other renewable sources such as solar and wind power, landfill gas feeds the electrical grid without interruption, no matter the weather or time of day, making it a reliable source.

Israel and his team have discovered that soaking leachate water into the landfill trash dramatically accelerates decomposition and landfill gas production. He hopes to expand Kiefer's system of underground infiltration trenches to wet more of the trash.

Speeding up the natural breakdown of buried trash also shrinks the volume of the landfill. "That extra airspace is money in the bank," Israel said. "The value of a landfill is in empty space, room for more waste."

E-waste recycling

At e-waste collection sites, printers, TVs, mobile phones, VCRs, laptops, and video game consoles are piled high.

There's gold in them thar hills. And platinum, copper, cadmium, mercury…

Electronic waste, or e-waste, is a type of trash that didn't exist decades ago when the county's Kiefer Landfill started operations. Now, e-waste is the fastest growing part of the county's waste stream, with over twenty-five million pounds processed in 2014, and increasing by over 10% a year. We own more electronic devices than ever, and we're rejecting or upgrading them with astonishing haste.

E-waste, which includes anything that has a circuit board **and** a power cord, is both a menace and an opportunity. Menace, because circuit boards and screens contain toxic heavy metals that if dumped in a landfill can poison the surrounding land and water. Opportunity, because e-waste also contains precious metals, aluminum, and high-quality plastics that are valuable if salvaged and recycled properly.

Because of toxic lead in the glass, in 2001 California banned landfill disposal of CRTs (cathode ray tubes), those fat TVs and computer monitors that used to swallow your desk. The only safe way to dispose of such screens is to disassemble them by hand, piece by piece. Because this is costly, vast quantities of American e-waste have been shipped to third-world countries by unscrupulous exporters who turn a blind eye to the fate of these devices. In the slums of China and Nigeria, desperately poor people have plucked apart our e-waste

Science in the Neighborhood: Utilities

under appalling conditions, using improvised smelters in huge trash dumps that quickly became some of the most poisonous environments on the planet.

In response to this travesty, California passed a bill (effective 2005) that required consumers to pay an electronic waste recycling fee, collected by the retailer, when they purchase any device with a screen larger than four inches. This pot of money is used to pay California recyclers for environmentally responsible dismantling of those devices here in the state.

For Sacramento engineer Paul Gao, this was a big opportunity. Gao had been recycling e-waste commercially since 2000. In 2003, his company CEAR (California Electronic Asset Recovery) contracted with Sacramento County to manage the county's e-waste, including CRTs. The new consumer fee greatly expanded his business and has assured that millions of pounds of discarded TVs and monitors were safely recycled close to home.

"I'm very proud that in 2010 CEAR was certified as an e-Steward by the Basel Action Network," Gao says. The Network is a global advocacy group dedicated to stopping the export of hazardous waste from rich countries to poor ones.

At Gao's Mather facility, devices with a screen larger than four inches, and printers/scanners with any kind of (mercury-containing) fluorescent light bulb, still must be taken apart by hand, a process subsidized by the recycling fee. Any batteries must be removed and are sent to a dedicated battery recycler. Dismantling all other "universal" e-waste is less labor-intensive. According to Kristin DiLallo Sherrill of CEAR, most e-waste recyclers use knife shredding: e-waste is cut into pieces by giant steel blades, followed by automated and manual sorting of the mixed fragments.

But Gao envisioned a better way. For the past five years, CEAR has operated a more efficient separator affectionately called "the green machine." This noisy green metal box the size of a two-story bedroom contains California's only centrifugal chain shredder. The green machine uses 30% less energy than a knife shredder, and produces cleaner output material, which has a higher value when sold to recyclers.

The green machine's main chamber spins a bit like a washing machine, but without plastering the e-waste against the sides. Instead, it hurls a pair of chains (fixed to the bottom) in such a way that a tornado-like vortex is created. This forces the e-waste up and into the air, where pieces collide against each other hard enough to crack the material into its component parts. This generates a lot of heat, so the chamber's walls are filled with water for cooling. The roof of the green machine is also covered by an airbag-like device as insurance against the slim possibility of an explosion due to a buildup of magnesium residue.

After the spin, broken fragments are separated by a giant magnet into ferrous (steel) and non-ferrous components. Then an eddy current device pulls out the aluminum. Humans sort the rest, picking out wires and plastics. The last one-third or so goes to a shredder. The resulting lower-purity bits are separated by color and shape by an optical sorting machine.

On a conveyor belt running out of the green machine, I saw weirdly intact blocks of aluminum, and metallic cylinders amid general brokenness. "Heat sinks," DiLallo Sherrill said. "Almost pure aluminum. And those are capacitors. Also mostly aluminum. If you run them through a knife shredder, the oils inside contaminate the rest of the material. But in the centrifuge, they stay whole." Elsewhere, a worker was shoveling green, metallic shards onto a conveyor belt. "Circuit boards," she said.

I asked DiLallo Sherrill about data security. If I turn in an old computer, what happens to the hard drive? "Any hard drives we get from the public, we shred," she said. Can't some be erased and re-used? "In theory, but we salvage drives only for certain specific customers' contracts."

Ultimately, materials are sold to recyclers who smelt the metals or re-use the high-grade plastics. Viability of the business depends on values in the market for metals, and also petroleum prices. Nothing toxic goes to the landfill. To responsibly get rid of your old electronics, never put them in the trash. Surrender them at a County drop-off center, or directly to CEAR, or donate them to an e-waste drive for charity.

Part Four

Ecology

'Tis the season for salmon

The first time I visited the Nimbus Hatchery on the American River during the salmon run, I thought I was in a nature documentary. So many Sacramentans are totally unaware of this annual spectacle and drama unfolding in the quiet waters near our homes. Seeing those big, beautiful fish jumping connected me to something primal and wonderful. Usually a good time to visit the fish ladder is right around Thanksgiving. A must-do for everyone, but especially for kids.

This article ran in December.

The American River is home to two native species of migratory fish that have the unusual ability to survive in both fresh water and salty ocean water at different times of their lives. These *anadromous* fish, steelhead trout and Chinook salmon, are hatched in rivers but spend their adult lives in the Pacific. After two to five years of life in the open sea, American River salmon return home to spawn, swimming 131 miles from San Francisco Bay to Sacramento.

The salmon start to arrive here about mid-October and continue into December, followed by steelhead in January and February. These fish will struggle upstream and uphill, leaping over rocks and obstacles, always fighting the current, committed to reaching the exact river and stream where they were born, navigating by sense of smell and magnetic fields.

Take a walk along the American River (especially upstream of Watt Avenue) and you may spot some of these magnificent fish swimming by. They might not look so magnificent, though. Salmon stop eating once they leave the ocean, and their bodies change shape for the spawning run. The difficulty of the journey is written in their ragged appearance.

In fact, the salmon are starving to death. This is a one-way trip. Their goal is to find a place with cold, clean water and a gravel bottom to build a *redd* (nest) where they'll lay thousands of eggs or fertilize them. Once that task is finished, both male and female salmon die. So fish carcasses might be easier to find than actual fish on your river walk. The odor of fish decay might reach you even if you're not at the river.

Migratory steelhead trout, on the other hand, do not exhaust themselves. They spawn and then swim back to the sea. An adult steelhead can make the long journey several times in its life. (Steelhead, by the way, are genetically the same as rainbow trout, fish that live entirely in fresh water.)

Prior to the Gold Rush, Chinook salmon were abundant in the American River. Although we can only estimate the numbers of fish from that time, it's clear that hydraulic mining and the construction of dams decimated their population. The American River used to have a second salmon run in the spring; due to human activities this run went extinct around 1950.

Dams cut migratory fish off from the majority of their historical spawning habitat, which once included about 6,000 miles of rivers and streams far up into the foothills. Now only one river (the Cosumnes) flows westward out of the Sierra Nevada with no major dam, leaving fewer than 300 miles of breeding grounds still accessible.

To compensate for the harm done by dams, the California Department of Fish and Wildlife operates ten salmon and steelhead hatcheries across the north state. Here in Sacramento, the Nimbus Fish Hatchery sits on the American River just downstream of Nimbus Dam near Hazel Avenue.

During salmon season, a removable barrier called a *weir* is placed across the river. This directs the fish into a fish ladder leading to the hatchery. If you've never seen the fish ladder in action, you simply must make a visit. Driven by their instinct to swim upstream, powerful salmon launch themselves uphill, one step at a time. For most of November and December, egg-taking operations are underway at the hatchery. Salmon are anesthetized in the water and sorted by sex, then quickly killed. Eggs are harvested from the females and blended with sperm "milked" from the males. Visitors can watch the whole process through windows in the visitor center.

Most of the fish returning in January are steelhead. Because they do not die after spawning in the wild, the hatchery breeds them without harm. The fish are returned to the river by a "slip and slide" tube down to the water. Baby steelhead are sheltered and fed for a year before release.

Bright orange, fertilized salmon eggs are stored in incubation jars. Tiny *alevin* hatch and grow into inch-long *fry* using the food stored in their yolk sacs. As they grow, the young salmon are transferred to larger tanks and to the raceways outside. Visitors are welcome to feed the fish in these long, seething ponds that surround the visitor center under netting to protect the fish from predatory birds.

At about six months of age, the small (4-5 inches) salmon are pumped into trucks and driven to release sites. Most are put into the American River in the Sunrise area. This proximity to the hatchery makes it more likely they won't "stray" when it's time to spawn; they'll come back to the hatchery. Survival rates are better, however, with fish released much further downstream in the Carquinez Strait, so some releases are done there.

Tens of thousands of pounds of edible salmon meat are donated by the hatchery every year and after processing are returned to local food banks. Lower quality salmon flesh is turned into fertilizer. If you'd like to catch some of that salmon for your own dinner, be aware that fishing is temporarily prohibited in the prime spawning habitat between Hazel Avenue and Ancil Hoffman Park. Downstream, you're welcome to try. Because the salmon are spawning, the quality of the meat varies. Fish who made the run up from the ocean quickly will be edible. Those who've spent more time in the river will be mushy.

Even if you don't fish the American River, you may have eaten a salmon hatched at Nimbus. The Nimbus Hatchery produces about 4,000,000 Chinook salmon and 430,000 steelhead trout every year. After being released into the river or downstream in San Pablo Bay, hatchlings that survive to adulthood join the wild populations in the ocean, where commercial fisheries catch fish destined for local supermarkets.

Salmon have another mammalian predator besides humans: sea lions. Remarkably, sea lions will follow salmon on the spawning run—all the way to Sacramento. According to Laura Drath, a Fish and Wildlife Interpreter at the Nimbus Visitor Center, sea lions are occasionally spotted as far upstream as Watt Avenue. She has seen them several times at Sutter's Landing (near CalExpo/East Sacramento).

The Nimbus Fish Hatchery and Visitor Center are open year-round. Call (916) 358-2884 for information about when the fish ladder is open and egg-taking is happening (weekdays only, usually Monday and Thursday mornings). Bring quarters for fish food.

Sacramento Splash spreads enthusiasm for threatened habitat

Nature conservation is an important issue for me. When I lived in Missouri, I supported organizations that protect native prairie. Here in Sacramento, I heard about a unique local habitat for the first time. **Vernal pools** are precious, irreplaceable, and disappearing. Unlike a redwood forest that automatically commands respect, vernal pools are easily dismissed as mere puddles. Splash's mission is to show citizens and policymakers there's more here than meets the eye.

A riddle for you: what am I?

Bees without hives
A bathtub with no drain
Flowers mark the end of lives
Dry shrimp will swim again.

This could be a real stumper. But if you're one of the

thousands of elementary school children who visited Sacramento Splash in the past year, you might have guessed the answer right away: vernal pools.

Vernal pools are an extraordinary Central Valley habitat that not enough people have heard of. Eva Butler, the founder of Splash, says, "Fifteen or twenty years ago, most people hadn't even heard of vernal pools. Part of the impact of Splash is now more Sacramentans know about them."

What is a vernal pool? The name gives us a hint. "Vernal" means spring. The most basic description of a vernal pool is a temporary pool of water that appears during our rainy season (winter and into spring), then dries up during summer and fall. But vernal pools are far more than simple puddles.

Vernal pools only form in places lined by *hardpan*, a layer of clay a few inches to a few feet beneath the surface of the ground that is so dense, it acts like a bathtub with no drain. Unlike a puddle, vernal pools don't drain away. Water leaves the pool only by evaporation, so pools linger until late spring or summer.

During the wet phase, vernal pools come alive. They are home to a wondrous array of animals and plants uniquely adapted to hatch, feed, breed, and die during this brief period. On Splash field trips, children (and adults) are amazed by what they see in a scoop of vernal pool water—an abundance of exotic creatures that resemble tiny aliens, among others, fairy shrimp (a relative of "sea monkeys" which some of you may remember from childhood); seed shrimp; clam shrimp; dragonfly larvae; and an endangered species found only in the Central Valley, the Vernal Pool Tadpole Shrimp.

Interestingly, you won't see many mosquito larvae in a healthy vernal pool ecosystem. Unlike a bucket of stagnant water on the side of your house, a vernal pool is loaded with predators that eat mosquito larvae and compete with them for food.

As the water dries up, the swimming creatures leave their eggs or cysts to wait for next year, and vernal pools put on their showy spring finery. It's the flowering phase! Primarily in April, the pools turn into muddy ground from which dazzling vernal pool flowers blossom. Species of solitary bees emerge from underground nests to collect pollen from a single kind of vernal pool flower upon which they depend for survival.

Splash organizes guided walks of the vernal pool flowers on Sundays in April. Reservations are required; visit SacSplash.org to sign up. The abundance and timing of the flowers' bloom are unpredictable, but Butler asserts, "No one ever leaves a Mather Field flower walk unsatisfied."

Sadly, this singular natural wonder is in danger of disappearing forever. Thriving vernal pools are part of a native California prairie (grassland) habitat. Prairie habitat has been utterly devastated across the Central Valley, with less than 10% of it left, a victim of urban development and agriculture. Sacramento retains two of the finest remaining vernal pool areas in the state (around Mather Field and Rancho Seco), but our county continues to lose thousands of acres a year of vernal pool prairie wild spaces. According to the Environmental Council of Sacramento, the County Board of Supervisors' recent approval of the Cordova Hills development outside Rancho Cordova will destroy some of the finest remaining pools in the Sacramento Valley. In the opinion of Splash Founder Eva Butler, "There is no local government commitment to vernal pool conservation in Sacramento."

In addition to suburban construction, the conversion of rangeland into vineyards is another threat to local prairie. "Agricultural operations {such as vineyards} bypass a lot of regulations that normally protect vernal pools," Butler says. "It's legal to plant grapes right around a vernal pool. Technically the wetland isn't filled in, but the prairie habitat doesn't function any more."

Sacramento Splash is dedicated to spreading the word about these local treasures. From their educational facility near Mather Field amid numerous vernal pools, Splash workers and volunteers lead tours, host schoolchildren, and maintain a website, www.SacSplash.org , that is the finest vernal pool education resource on the Internet. "If you take young people to explore the place they live, they let their families know there's something special to see here," Butler explains. "Our home-grown habitat isn't impressive on the scale of mountains or redwoods, but on a small scale our vernal pools are spectacularly complex and beautiful."

"You don't have to go somewhere else to see nature."

Citizen scientists help save our elms

Sacramento is known as a City of Trees thanks to ongoing efforts by government, private agencies, and citizens. Do you appreciate our urban forest? Ask SMUD about their free shade tree program, or read on to learn about a special citizen-science project with the Sacramento Tree Foundation.

Sacramento became a "city of trees" thanks to 19th century residents who, desperate for shade, planted elms—graceful, fast-growing, long-lived trees ideal for city life. Tens of thousands of elms able to grow over 100 feet tall kept the city cooler in the summer and provided a beautiful canopy that still defines Sacramento's older neighborhoods.

Today the magnificent elms of our urban forest are at

risk. According to Pamela Sanchez, a certified arborist with the Sacramento Tree Foundation, Sacramento may have had 25,000 elm trees in the past. Now we're down to about 2,000. Many elms have died of old age and were replaced with a different type of tree, or not replaced at all. Others were casualties of major storms in the mid-twentieth century. But starting in 1990, Sacramento's stately elm trees faced a deadly new threat: Dutch elm disease (DED).

DED is caused by a fungus that probably originated in Asia. Both European and American elm species are susceptible. DED came to North America from Europe via infected logs around 1930. (The modifier "Dutch" refers to a group of scientists who studied the fungus, not where it came from.) The disease marched across the continent, slaying tens of millions of trees, and reached Sacramento in 1990. The fungus infects the tree's water-conducting (vascular) system. The tree tries to protect itself by blocking off the affected tissue, but this clogs the tree's "pipes." Water can't reach the crown, and the tree dies.

Elms catch DED in two ways: bark beetles and root grafts. Bark beetles tunnel into elm wood under the bark and lay their eggs. The eggs hatch, the larvae mature into adults, and adult beetles exit the tree. If the tree is infected with DED, spores from the fungus stick to the beetles who then carry the disease to healthy elms. Trees infected in this manner will show wilting, curling, or browning of the leaves on the infected branch. The leaves will often drop prematurely. The disease will steadily progress through the entire tree, one branch after another. The elm can survive for several years before perishing.

The fungus can also spread via the roots. Elm trees planted in a row along a city street, for example, have roots that cross each other in the soil. Eventually the roots grow together, or "graft." This means the trees now share a vascular system, and if the DED fungus is in one tree, it can flow into other trees in the row too. Elms infected via root grafts will die rapidly (weeks to months rather than years), causing neighborhoods to lose entire stretches of big old trees at once. This has been a lesson to urban foresters. Sanchez says, "There used to be less variety in the urban forest. Now we try to manage away from monoculture {planting only one kind of tree} on a single street."

Dutch elm disease is incurable so the goal of management is to prevent transmission. Once a tree is infected, its root grafts to neighboring trees must be severed and the tree must be cut down as soon as possible, before the bark beetles emerge. In the 1990s when many elms were dying, Sacramento launched an early detection program. The program itself died of budget cuts but was resurrected in 2016 by the Sacramento Tree Foundation and the City of Sacramento. The new Save the Elms Program (STEP) relies on citizen-scientist volunteers to monitor public elm trees. These trained volunteers visually inspect their assigned trees at least three times each summer, looking for signs of ill health.

According to Matthew Van Donsel of the Sacramento Tree Foundation, in this first year they trained 50 volunteers who monitored a total of 750 elms. "At the training session we walk people through how to identify English and American elms. We teach them how to identify symptoms of Dutch elm disease, and generally help

Science in the Neighborhood: Ecology

people get used to looking at trees." Citizen-scientists record their observations in an app called Greenprint-Maps, which has records of all the city's public elms. They use binoculars to examine the tree's crown, and if they see anything amiss, they take a photo and alert the Foundation. The Foundation passes the information to the City's Urban Forestry Department for follow up.

"It's hard to tell the difference between drought, DED, and squirrel damage," Van Donsel admits, "but people get better as they do it." The more eyes on the elms, the better. "If you live in Midtown or downtown and have an elm on your street, look at your tree every few weeks, especially in the summertime when DED shows itself. Take a walk, save a tree."

The most important thing you can do to keep your elm tree healthy is water it deeply in the summer. Avoid unnecessary pruning of elms in the spring or summer because at that time of year the "wounds" may attract bark beetles. Chemical fungicides can suppress DED spread by bark beetles, but the treatments are expensive, unreliable, and can shorten the tree's lifespan so the City does not use them on the general elm population. "Some private residents of Curtis Park have elected to pay for fungicides themselves," Sanchez says. "It's worth it to them to try to protect their trees."

Sanchez argues that saving Sacramento's elms is worth a lot. "Their significance can't be overstated. They provide many benefits: shade, beauty, cleaning the air—they're not replaceable."

Training sessions for new citizen-scientists are held in spring. Learn more at sactree.com/STEP.

Citizen scientists tally species in the Great Backyard Bird Count

> Birds are a terrific way to introduce people to nature and ecology. You don't have to travel to see them. In fact, you don't even have to go outside. Sacramento yards and urban parks have many birds, including big charismatic ones like hawks and vultures and turkeys, colorful ones like scrub jays and yellow-billed magpies and Anna's hummingbirds. Can't tell a swallow from a sparrow? Start by identifying one bird a week from a window in your home.

When I lived in Minnesota, we envied the birds that flew south before the onset of our frigid winter. The first robin to return home was always a happy harbinger of spring. This explains my particular delight here in Sacramento when I see an entire flock of red-breasted birds—in *January*.

But there's more than robins in my yard. Have you seen the birds in Sacramento in winter? You probably notice crows or pigeons, but what about the birds that flit past the corner of your eye, that roost in the trees or hide in the shrubbery, or gather in the water around the Yolo Causeway? Have you ever taken a few minutes to really look at them?

You should. Sacramento is a hub of bird activity in the winter. Tens of thousands of birds *come* to town, rather than leave, drawn by our mild temperatures and wetlands. These avian tourists arrive from as far away as the Arctic Circle, traveling a superhighway in the sky known as the Pacific Flyway. The flyway stretches all the way from Alaska to South America. Like any freeway, it has rest stops along the way, places where weary birds can eat and regain their strength.

The Sacramento region is a major rest stop, or even winter home, for migratory birds. The Yolo Basin Wildlife Area around Interstate 80 between Sacramento and Davis hosts thousands of traveling birds, including many species of ducks that feast on leftover rice in the fields, and high-flying snow geese, white with eye-catching black wing tips. The Cosumnes River Preserve (south of Elk Grove just off I-5) is noted for its sandhill cranes, impressive birds with a six-foot wingspan and distinctive call. National Wildlife Refuges at Stone Lakes (south on I-5), Colusa (40 miles north of Woodland, off I-5), and Sacramento (another 25 miles north of Colusa) are bustling with waterfowl and other bird species in the winter.

Birding (the term aficionados prefer over "birdwatching") in these places requires nothing more than time. Part of the fun, though, is identifying particular species. There is a scavenger-hunt kind of thrill to checking birds off a list. It's easier than you might think. Just grab a pair of binoculars, print a page of bird pictures specific for our area (such as "What's This Bird?" at SacramentoAudubon.org), and head out. The visitor centers at Cosumnes and Yolo also offer bird guides. Or carry a smartphone app, which will give you photos and recorded birdsong to help with identification. Merlin Bird ID (free) and iBird (not free) are both great for beginners.

You don't have to leave your own neighborhood to identify remarkable birds. Jenner Junghans, Education Chair of the Sacramento Audubon Society, says, "When people who are not birders see photos of our local birds for the first time, they're often stunned. They had no idea we have local birds with such bright colors and bold patterns. People feel like they're looking at photos of birds from someplace like South America."

Some local birds are even international celebrities. Every day in my Arden neighborhood I see large, beautiful, black and white birds with long tail feathers and striking yellow beaks. This seemingly common bird is in fact a rarity. The yellow-billed magpie is an *endemic* species, meaning it's not found anywhere on Earth except California's Central Valley.

February is the perfect time to give birding a try. President's Day weekend is the Great Backyard Bird Count, a worldwide annual event that anyone can participate in. GBBC is an important scientific research project that relies on ordinary people to collect data. It's quite simple. For 15 minutes, count all the birds you see, whether you're gazing out your kitchen window, walking downtown, or hiking through a nature area. Submit your location and number of each species you saw to gbbc.birdcount.org. With data collected by a large number of people in a variety of places, scientists are able to track bird populations and migration patterns. Over time, this information helps scientists see trends and determine how bird populations are affected by development, habitat loss, disease, and climate change. Scientists can then make recommendations to conserve and protect habitat and birds, all because of the combined efforts of citizen scientists—people like you.

If you'd like to join in the GBBC but you don't know a meadowlark from a mockingbird, participate in a birding event for beginners with the Sacramento Audubon Society (SacramentoAudubon.org) or attend the annual Bird and Breakfast at Effie Yeaw Nature Center (SacNatureCenter.net).

If you buy just one bird guide, I recommend **Birds of Northern California** *by David Fix and Andy Bezener (Lone Pine Publishing)*

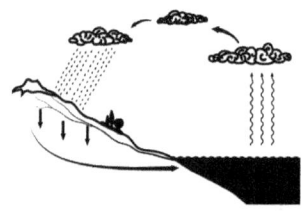

Nature Bowl engages kids in conservation

In the old days, kids intuitively knew their local environment and habitats because they played and worked in them. They saw the animals and plants, and the change of seasons. Nowadays, if most kids are to learn any of that, somebody has to lead them. Bruce Forman is a visionary environmental leader who created a program to lead kids into nature. I coached Nature Bowl teams for four years and it was perhaps the best thing I've ever done. Can you get involved?

Show a picture of an African animal like a zebra or giraffe to a child in Sacramento and likely they'll be able to tell you what it's called. But show them a photo of a local sandhill crane or coyote and few will know its name. Time that kids used to spend exploring nature outdoors is now more likely to be spent inside or on a sports field. Ironically, care for the environment as a value has been rising at the same time that actual understanding of local ecosystems has been falling. Children may feel a responsibility to protect endangered species of the rainforest even though they can't identify animals, birds, or plants that live right here.

For more than thirty years, Bruce Forman of the California Department of Fish and Wildlife has been working to change that. In 1986 he launched Nature Bowl, a program exclusively for 3rd-6th grade students in the Sacramento River valley that according to Forman "serves to increase science and conservation literacy with a focus on local environments."

Most Nature Bowl teams are coached by school science teachers who integrate the Nature Bowl curriculum into their regular classes. Parents, scout leaders, and others can organize teams, too. 90-95% of teams are from public schools. Some Sacramento schools have participated for decades. In Nature Bowl, students learn about local habitats—riparian, grassland/prairie, vernal pools, oak woodland, wetland, etc.—and the typical species that live within them. Field trips to local wilderness areas such as Effie Yeaw Nature Center and Cosumnes River Preserve can greatly enhance student interest and learning. Classroom study includes an impressive list of vocabulary words (do you know what *crepuscular* means?), and study of the principles of ecology such as food webs and energy pyramids.

Nature Bowl isn't just about learning information. It's also about engaging with real life. In one Nature Bowl event, students create a one-minute "enviro-mercial," which is a persuasive speech describing an environmental problem or issue specific to the Sacramento area. Forman strongly encourages students to not only make their presentation but to take some action on the topic, for example, to write a letter to the newspaper or a County supervisor, to go out and clean up trash, or to set up a compost bin at home.

Preparation culminates in team semifinal events held in March and April at a variety of outdoor settings in the region from Auburn to Davis. Kids love the outdoor Nature Investigations event, when they act as detectives searching for un-natural items, signs of wildlife, and living specimens. During the Bell Ringers event, students amaze the parents with their answers to questions that sometimes stump the audience. Forman emphasizes that although two teams from each semifinal site are chosen to advance to the finals in May, the events are structured to be noncompetitive, fun, and educational. Forman says, "Nature Bowl isn't just for advanced students. It's for all students. There's no ranking of the teams, and they all get the same prizes. Everyone is commended for what they've done." He wants every participant to walk away feeling like they did well. "I've heard kids say to their parents, 'We won! We won!' because they succeeded in a section, even if they didn't advance to the finals."

By several measures, Forman's creation has been a success. The number of kids involved has grown, teachers praise the effectiveness of the program, and Forman says he runs into Nature Bowl alumni almost every month. Some tell him it was one of the best parts of their school experience. Alumni in their twenties and thirties come back to volunteer at the finals, and some even have children of their own participating. Forman takes particular satisfaction in knowing that many Nature Bowl kids develop a passion for making sustainable choices in their personal lives. He's met some who have extended that passion into volunteerism and even careers in environmental activism.

If Nature Bowl is not available to your child, you can start and coach a team. Too intimidating? Maybe your child's science teacher has not heard about the program, or is too busy to manage it alone. If you offer to co-coach, perhaps you can bring this life-changing experience to your child and others.

Forman summarizes the value of the program. "Especially with the advance of technology, Nature Bowl provides a window to the outdoors that students need more than ever. It energizes the kids not just academically but more holistically to get engaged with the natural environment and really like it."

To learn more about starting a team, attend a coaches' workshop for teachers and parents in January or February. Get coaching materials, walk through Nature Bowl activities and hear sample questions. For dates and locations, visit wildlife.ca.gov/regions/2/nature-bowl or call Joanie Cahill at (916) 358-2852.

Fight the bite: Mosquito and vector control

I grew up in Minnesota, where we like to say the mosquito is the state bird. In my small rural town we certainly didn't have an agency tasked with trying to control the mosquito population. It showed. Red, itchy welts were just another sign of summer.

In a 2014 blog post, Bill Gates asked, "What is the most dangerous animal on Earth?" Defined as the one that kills the most humans, the answer might surprise you: it's mosquitoes.

Mosquitoes cause the deaths of roughly a million people each year (most from malaria) because they're *vectors*. A vector is an insect or animal that transmits a disease to other animals or humans. In Sacramento, mosquitoes spread West Nile virus, Western Equine Encephalomyelitis virus, St. Louis Encephalitis virus, and canine heartworm. In the early 1900s, malaria was also a serious local threat. Thanks to mosquito control efforts, the only malaria cases in our area occur in travelers returning from foreign lands. Zika transmission is not a concern at present; the types of mosquitoes that can carry Zika virus are not found in Sacramento or Yolo County. But new mosquito-borne viruses like Zika could emerge at any time.

Mosquitoes are therefore more than a nuisance. They're a threat to public health. In 1946, a mosquito abatement district was created by voters in the Sacramento region. Later, the renamed Sacramento-Yolo Mosquito and Vector Control District (MVCD) added ticks (vectors for Lyme disease) and wasps to its services. Like other California independent special districts, MVCD pools resources across city and county lines to provide a focused, essential service. It's governed by a twelve-member Board of Trustees appointed by each of the counties and incorporated cities in its jurisdiction. The district is funded by property taxes and has specific powers authorized by law.

MVCD's main job is to monitor mosquitoes and their diseases, and to take actions to "fight the bite"—to minimize mosquito populations and their impacts on people. Six of over twenty local species of mosquitoes are significant for public health. Each carries different diseases and has different feeding and breeding habits. Some species are most active in summer and fall, others in late winter through early spring. Some prefer to feed on mammals, others on birds. Some lay eggs in rice fields, others in tree holes, and another in foul water such as drainage basins. For all mosquitoes, the life cycle is the same. An adult lays a raft of eggs on the surface of water. Larvae hatch and turn into pupae just under the water's surface. Adults emerge and fly away in as little as a week.

Mosquitoes will breed just about anywhere that still water collects. Agricultural sites (especially rice fields) and stormwater systems can yield a lot of mosquitoes. Urban sources include fish ponds, unmaintained swimming pools, containers, bird baths, blocked roof gutters, irrigated lawns, cemetery vases, pet dishes, and abandoned tires.

As part of a multi-pronged approach called Integrated Pest Management, MVCD educates people about their role in mosquito control. The district encourages actions that deprive mosquitoes of breeding habitat, and actions to prevent mosquito bites which spread disease.

Here are their recommendations to the public, summarized as the "Seven D's": **Drain** any standing water on your property. **Dawn** and **dusk** are when mosquitoes are most active, so stay indoors. When outdoors, **dress** appropriately with long sleeves and pants. **Defend** yourself with insect repellent. Check **door** and window screens; repair any holes. If you have a problem or question, call the **district**.

MVCD uses science-based surveillance and testing programs to measure vector activity. The district observes mosquito populations over time in all habitats of Sacramento and Yolo counties. They use traps with carbon dioxide as a lure to catch and count adult females, the only mosquitoes that bite and take a blood meal. (Males innocently dine on flower nectar.) Weekly mosquito counts are posted online. Collected females are also tested for viruses to see if there is disease in the population, and how prevalent it is.

In addition to looking directly at local mosquitoes, the district monitors birds, which are good indicators of mosquito-borne disease activity. Crows, jays, and magpies are particularly sensitive to West Nile virus, so MVCD collects dead birds and sends them to a UC Davis laboratory for testing. Dead bird reports from the public are particularly important early in the season. If you find a dead bird, call 877-WNV-BIRD. Likewise, the district surveys veterinarians for data about heartworm cases in dogs, and determines areas of higher risk for heartworm transmission.

Furthermore, MVCD maintains four "sentinel" flocks of five chickens each at strategic locations in the area. If bitten by mosquitoes carrying a virus, the chickens do not get sick but they will develop antibodies against the virus. Blood samples are taken from the chickens every other week during mosquito season (May to October) and tested for those antibodies.

When surveillance indicates that mosquitoes are spreading disease in the region, what actions does the Mosquito and Vector Control District take?

Killing Mosquitoes

The Sacramento-Yolo Mosquito and Vector Control District (MVCD) protects public health by minimizing the spread of infectious diseases carried by mosquitoes and other insect vectors such as ticks. Many local residents know the district for only one thing: mosquito spraying. But chemical spraying to kill adult mosquitoes is one tool in a multi-pronged strategy called Integrated Pest Management. The primary goal of this strategy is to prevent mosquitoes from maturing in the first place.

All species of mosquitoes hatch their eggs in water. Without suitable water sources, mosquitoes can't breed. Therefore managing water sources through education and enforcement is a big part of what MVCD does. For the general public, the main message is to empty any standing water on your property. Farmers are advised to adopt agricultural practices that minimize mosquito breeding, such as proper timing and methods to flood or irrigate a field.

After education, the next tier in MVCD's Integrated Pest Management is *physical control*. The district has projects to improve rural drainage, clear vegetation and sediment from channels, and do construction to eliminate standing water.

Biological control is the use of natural predators to eat mosquito larvae in water sources that can't be eliminated. According to Luz Maria Robles, Public Information Officer for the district, MVCD has 23 ponds in Elk Grove where they breed mosquitofish (*Gambusia affinis*). Mosquitofish are an ideal biological control method. These fish are small (about 1-2 inches), they breed quickly when established in a water source, and they survive Sacramento winters. Each fish consumes 200-300 mosquito larvae per day. They require minimal care and can survive in a variety of environments such as rice fields, pastures, fountains, ponds, abandoned swimming pools, even animal watering troughs. The district produces over 4000 pounds of mosquitofish per year and gives them free to the public—all you have to do is ask. They will even deliver the fish to you. MVCD also stocks guppies (*Poecilia reticulata*), which are useful in polluted water sources and low-oxygen environments. Guppies do an excellent job of controlling mosquito larvae during the summer months but do not survive the cold.

The district also uses a method of *microbial control* to kill mosquito larvae: a naturally occurring soil bacterium called *Bacillus thuringiensis israelensis* (Bti). Bti bacteria can be applied to water. Mosquito larvae then eat the bacteria's spores, which carry natural bacterial toxins into the mosquito's gut and ultimately kill it. Bti is used all over the world because unlike most chemical insecticides, its toxicity is very specific to mosquito and blackfly larvae. Other insects and organisms are not affected.

Finally, *chemical control* is a key part of Integrated Pest Management. MVCD does both agricultural and urban spraying of insecticides to reduce adult mosquito populations when circumstances require. All products used are registered with the California Environmental Protection Agency. Because the effectiveness of various chemicals can decrease over time, part of MVCD's job is to regularly test local mosquitoes for resistance to insecticides.

Chemical spraying is done only when data from MVCD's surveillance indicate a public health threat. Warning signs include finding virus-positive dead birds or mosquitoes; an increase in mosquito abundance; or a rising infection rate in mosquitoes. These signs of a "hot spot" prompt additional testing, as do human cases of mosquito-borne diseases such as malaria or dengue that are reported to MVCD by the public health department. The district follows up with additional mosquito trapping in the vicinity of where the (unidentified) person lives.

Because the decision to spray is data-driven, there is no planned schedule for spraying. Anyone who wants advance notice of spraying should sign up for the district's email notification list at fightthebite.net/spray-notification/ or watch the updates at fightthebite.net/spraying-update/. Every year MVCD conducts aerial spraying of agricultural areas such as rice fields and pastures. Urban spraying is calibrated to match the size of the area where action is needed. Ground spraying is done using backpack foggers, hand sprayers, or trucks. If the area is too large for these methods, chemical sprays are delivered by aircraft. Environmental conditions such as temperature and wind speed must be suitable for spraying to take place. The insecticides break down in sunlight, and are said to only kill mosquitoes and things smaller than mosquitoes. To further protect "good" insects, spraying is done at dawn and dusk.

"It's all an effort to prevent human cases," Robles says. Not everyone is happy about this part of MVCD's work, but the special district has legal authority to overrule objections to urban aerial spraying such as those raised by the Davis city council in 2014.

What would Sacramento be like without the district's efforts? With mosquito populations unchecked, it's possible malaria could come back. At the least, Robles says, "People would certainly be complaining about mosquitoes!"

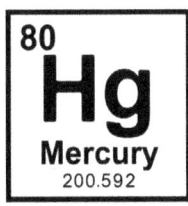

Mercury: A toxic legacy of greed

In addition to my freelance science journalism, I write and review science-themed novels at ScienceThrillers. com. An indie book about a team of forensic geologists working in the Sierra Nevada came to me for review.
Forensic geology—who knew such a marvelous specialty existed? Quicksilver by Toni Dwiggins also got me thinking about how much mercury escaped in the mountains during the heedless quest for gold...

The California Gold Rush brought fame, growth, and prosperity to Sacramento. It also destroyed a way of life for the Native Americans, and left environmental damage that impacts us today. Perhaps the most significant environmental problem left behind is mercury contamination of our local waterways.

Mercury (abbreviation: Hg) is a neurotoxic metal that occurs naturally in several different chemical forms.

California's Coast Ranges are rich in ores of mercury sulfide (HgS), in a mineral called cinnabar. Cinnabar is converted into elemental mercury (just plain Hg) by heating in a furnace. The mercury vaporizes and is condensed into the silvery liquid you might recognize from old thermometers. California cinnabar mines have produced a lot of mercury—more than 200 million pounds since 1848.

More than 25 million pounds of that mercury was used for gold recovery by mining operations in the Sierra Nevada and Klamath-Trinity Mountains, primarily for placer mining. In placer mining, flakes of gold are extracted from large amounts of sand or gravel tumbled with water over some kind of sluice box. Because gold is more dense than other substances in the slurry, it falls out of solution into the box's riffles. Some gold, however, is lost with the waste water and rock.

Elemental mercury makes placer gold recovery more efficient because gold and mercury stick together. Miners would add mercury to their sluice boxes, capturing bits of gold as a heavy amalgam of the two metals. The amalgam would be trapped, and they would heat it (a process called retorting) to vaporize the mercury and purify the gold.

Although miners attempted to trap and reuse their mercury, an estimated 10-30% was lost into the environment—millions of pounds of mercury. Some went into the air; most entered the water. Studies are ongoing to understand where all that toxic metal is today. Some remains in soils and bedrock at mining sites. Hydraulic mining sites, where mountains of earth were blasted apart and run through sluices, left behind huge amounts of mercury--according to a US Geological Survey fact

Science in the Neighborhood: Ecology

sheet, hundreds to thousands of pounds remain at each site. Much mercury was swept downstream with placer tailings and lies in sediments which continue to hold mercury today.

As a result, several local waterways with a history of gold mining are now on the Clean Water Act Section 303(d) list of "impaired" waters due to mercury levels that exceed water quality standards. These include the American River (below and above Nimbus Dam), Sacramento River, Feather and Yuba Rivers, Cache Creek, Davis Creek, Putah Creek, Folsom Lake, and Lake Natoma. Actual concentrations of mercury in these waterways varies a lot. Generally there's more mercury present in areas immediately downstream of old gold or mercury mines, and more mercury after flooding or high streamflow.

By itself, tiny concentrations of elemental mercury does not pose a threat to human health. For example, it's perfectly safe to swim in these waters. Elemental mercury has a hard time entering the human body. It's not absorbed through the skin, and even if consumed, very little of it gets absorbed by the gut. (The elemental form is hazardous primarily through inhalation of its vapor.)

Unfortunately, elemental mercury in the environment doesn't stay in that chemical form. Under low-oxygen conditions (such as in sluggish water or wetlands), bacteria convert Hg into methylmercury. Methylmercury is the most toxic form of the element. When ingested, methylmercury is readily absorbed into the body, where it damages the nervous system, immune system, and more. Unborn babies and infants are exceptionally sensitive to the toxic effects of methylmercury. Exposure can cause birth defects, or symptoms that are hidden for a long time.

So how might you be exposed to methylmercury? By eating fish from impaired waters. Fish gradually accumulate methylmercury from the water they live in. It enters their muscle and liver. If one fish eats another fish, the predator absorbs the mercury from the prey. This means mercury concentrates up the food chain. Top predators like bass (as well as fish-eating mammals and birds like eagles) show the highest concentrations of methylmercury in their tissues, a phenomenon called biomagnification.

Humans who eat contaminated fish are in the same boat as other top predators. The more they eat, the more mercury they accumulate in their bodies. Unlike some organic toxins found in fish, mercury can't be reduced by the method of preparation of the fish. It's in the fillets themselves, and cooking doesn't eliminate it.

The solution is to pay attention to fish advisories and avoid eating large amounts of fish from contaminated waterways. What exactly does that mean? California's Office of Environmental Health Hazard Assessment (OEHHA) posts recommendations of how much fish is safe to eat based on the fish species, the waterway it came from, and who you are (child, woman of childbearing age, or not). For example, young women are advised that lower American River steelhead trout are safe to eat twice a week, but black bass from the same river should be entirely avoided.

Find fish advisories statewide at oehha.ca.gov/fish/advisories.

California Naturalist Program fosters stewardship

My daughter always loved nature. As a teen she volunteered at Effie Yeaw Nature Center, and that's where we learned about a formal certification program for people who want to be educators, advocates, and investigators for California ecology. She completed the course, and takes her nature journal wherever she goes.

California is a place of extraordinary ecologic diversity. Habitats in our state range from frigid high-altitude Sierra slopes to the sizzling below-sea-level expanse of Death Valley, with grasslands, rivers, intertidal areas, and forests in between. These many habitats host complex ecosystems of plants, animals, insects, birds, microorganisms, and forty million people. Those forty million of us have altered the ecosystems to serve our needs. The changes we've made to our natural world affect our water quality, air quality, food production, and health. Are we making the right choices?

That's a question of both science and values. Scientists in a variety of disciplines focus on the natural world and the effect of human activity. But should this be a question left only to professional scientists?

Californians with a passion for the natural world can get involved in stewardship of our resources by becoming a certified California naturalist. The California Naturalist program is designed by the University of California and delivered by local environmental organizations. Participating adults study California's unique ecology in a 40+ hour course that's much more than just a science class. While there is a textbook and some lecture time, field experience, problem-solving, and community service are core parts of the curriculum.

The goal is to create a cadre of people with the knowledge and communication skills to participate in all kinds of citizen-level activities to help protect and restore California nature. According to the program organizers, certified CA Naturalists participate in scientific research such as plant identification; environmental monitoring such as bird counts and invasive species watch; habitat restoration or conservation such as seed collection; and importantly, education to make science more accessible to others.

The American River Natural History Association is one of the organizations that run the course around the state. ARNHA's group meets at Effie Yeaw Nature Center in Carmichael, and in the fall of 2017 about twenty Sacramentans were working toward certification.

Science in the Neighborhood: Ecology

Their instructor embodies the values of the naturalist program. Mike Cardwell wasn't a professional scientist. He'd worked in southern California law enforcement for decades when he completed a master's degree in ecology. Since then he's gone from reading books to writing them and has spent lots of time out *in* nature, studying and engaging with wildlife.

What kind of wildlife? Rattlesnakes. This makes education and science communication an important part of what Cardwell does. "My primary interest and expertise has always been wildlife that others unreasonably fear, particularly venomous animals and the injuries they sometimes cause," he says. He's currently running a multi-year study of the natural history of rattlesnakes at Effie Yeaw. Using miniature surgically implanted radio transmitters, he can track these elusive creatures and fill in some of the huge gaps in our understanding of how they actually live. "It's difficult to dispel myths and folklore without credible, well-documented sources of alternative information," Cardwell says.

Students in Cardwell's California Naturalist class learn how to keep field notebooks and participate in a capstone project that bridges the naturalist education of the course with community service or education. In the end they will join over 1500 Naturalist alumni of the certification program around the state and as Cardwell says, "part of an exciting and venerable tradition of naturalists throughout history."

If you'd like to become a certified California Naturalist, visit calnat.ucanr.edu. In addition to ARNHA, local course providers are the Yolo Basin Foundation in Davis; Tuleyome in Woodland; and the American River

Conservancy in Coloma. Want to make a vacation of it and get it done quickly? One week immersion courses are also available around the state.

Part Five

Community

Science Café Society

Drink, dine, discuss science: sounds like my kind of night on the town.

Raise your hand if you like to sit still in a lecture hall for an hour while a lecturer drones on using specialized vocabulary you don't understand. Sound like your memory of science class? Time to discover the new way to experience science: science cafés.

Science Cafés are a global movement to create informal, engaging conversations about science between an expert and a gathering of interested people who may or may not have expertise in science, all arranged in a casual, fun venue like a pub or coffeehouse. Think of it this way: instead of going to school, you sit down for a beer with your favorite teacher and just talk.

Like TED talks, the science café concept has really taken off. A lot of young people going into science these days appreciate the importance of scientific literacy in creating a climate of support for their work. They feel a responsibility to learn good science communication skills, and to apply those skills in their local communities. They're organizing science cafés like Café Scientifique in Silicon Valley, Nerd Nite in East Bay, Science Buzz in Sebastopol, and Science on Tap in Santa Cruz.

Here in the Sacramento region we've got a whopping THREE general interest science cafés, covering most Wednesday nights of the year. The most convenient for INSIDE readers is Sac Science Distilled, launched one year ago thanks to the efforts of local biology students Bobby Castagna and Nicole Soltis. In partnership with the Powerhouse Science Center, Capital Science Communicators, and Davis-based Science Says, Sac Science Distilled brings together scientists and the public at Streets Pub and Grub (1804 J St) every third Wednesday at 6 PM.

Co-founder Nicole Soltis says, "We started {the café} to give early career scientists—grad students, post docs—a chance to present their research to a general audience, and to practice talking about their work to the public." Recent topics included microbes in space, ringtails in the Sutter Buttes, and parasites. According to Soltis, Sac Science Distilled promises "short, idea-centered talks from local experts in the sciences" followed by "a lively discussion that brings science into context for everyday life."

I had to check this out.

The night I attended was a special event: the first annual Sac Science Idol, a speaker competition among a dozen young scientists who each had 3 minutes to wow the judges with their style and content. An enthusiastic audience filled the bar, swilling beers and cheering them on. The night's winner, A.J. Slepian, dressed up as an electron-carrier molecule (well, sort of) and convinced us to watch for news about NAD+ and aging. Attendees played Science Idol bingo with squares like "bad microphone feedback" and "cancer" added to the fun. Soltis says they usually have some kind of hands-on feature like that to engage the crowd. For example, after a recent talk about microbes the audience swabbed different locations in the bar and streaked petri dishes to see what would grow. At an animal behavior talk, each person got a card assigning them an animal identity. Then they had to act out a behavior that would help them find the right mate. A geek icebreaker and dating game all in one!

If you're willing to venture a little farther than Midtown, more science cafés await. Davis Science Café was started in 2012 by UC Davis chemist Dr. Jared Shaw. Shaw is still at the helm of this second-Wednesday gathering, currently at G Street Wunderbar in Davis. Lodi joined the fun in 2014 when Nick Gray, Education Program Coordinator at Lodi's WOW (World of Wonders) Science Museum, launched Science Night Live. SNL is held the first Wednesday of every month at 6 PM. Though it's held at the museum not a café, the event typically features a food truck and a cash bar. Gray calls

SNL "an engaging science experience within a casual atmosphere for the curious minds in our community." Along with conversations and demonstrations about robotics, black holes, consequences, groundwater, women in SciFi, bats, and alpacas, Gray has spiced things up by adding some science trivia nights too.

Looking for even MORE science cafés? Health-related topics are the focus at the Community Conversations Science Café hosted by UC Davis Research and Education Community Advisory Board, with support from Powerhouse Science Center. These events take place periodically on the fourth Thursday at Old Soul @ 40 Acres (3434 Broadway). Find schedules for this and the other science cafés at capscicomm.org/local-science-cafes/.

Sacramento Museum of Medical History tells stories of the past

> This place is one of the secret gems of our town. Tucked away in East Sac, it's well worth a visit. I loved seeing the iron lung but the museum's greatest treasure is its curator Dr. Bob LaPerriere.

Imagine going to the doctor and having leeches applied to your skin, or being given a medicine containing arsenic, strychnine, and mercury. Such practices were standard in Sacramento from the time of the Gold Rush into the 20th century. The Sierra-Sacramento Valley Medical Society tells the story of those days at its Museum of Medical History, located on Elvas Avenue in East Sacramento.

Dr. Bob LaPerriere, Curator, gave me a tour of the museum, which displays part of the Society's eclectic collection of healthcare-related objects from Sacramento's past. From macabre to artistic, the museum is a treasure house of interesting stuff. When you first enter the door, you encounter one of only a handful of fully functional iron lungs on public display in the US. This person-sized metal tube looks like a prop in a sci-fi space travel movie. In fact, it's a relic of a terrible time when polio left children paralyzed and fated to spend the rest of their lives immobile inside this clunky breathing machine.

Exhibits in the museum are brightly lit and organized by theme. LaPerriere told me stories as we went. At a collection of bottles from early Sacramento pharmacies, embossed with names like "AC Tufts, Druggist" and "Peters and Ray" and "Southern Pacific Co. Hospital Department," LaPerriere said, "Before the 1906 Federal Food and Drugs Act, these so-called patent medicines were often labeled 'cures.' After 1906, they had to call them 'remedies.'"

For those who relish gore, the museum offers an assortment of bloodletting devices designed to bleed a patient; Sadie, a human skeleton; plenty of primitive surgical instruments; and even a preserved jar of *Ascaris lumbridcoides*, a gigantic parasitic intestinal worm that was once common in the US. A certain beauty can be found in other artifacts. LaPerriere pointed out a plywood leg splint hanging on the wall. "Someone from the Crocker Art Museum was here and identified that as an Eames splint." (Charles and Ray Eames are famed American designers best known for their chairs.)

Sacramento medical history wouldn't be complete

without traditional Chinese medicine. Chinese herbal pharmacies have been in our area since the Gold Rush. The Yee family, a local clan with deep roots in Sacramento health care, has loaned an assortment of Chinese herbs and related items to display at the museum. You'll also see artifacts from noted Sacramento physicians of the 19th century, including a top hat and cane belonging to Alexander Butler Nixon, president of the California Medical Society who came to town in 1849.

The march toward modern, evidence-based medicine is evident in the collection. There's the city's first x-ray tube, from 1897, as well as early blood pressure measuring devices and ophthalmoscopes for visualizing the retina. I was fascinated by a series of artificial heart valve prototypes, developed locally in the 1960s by Dr. Edward Smeloff of the Sutter Research Institute and engineers at Sac State.

From the labels I noticed a number of items in the museum were donated by Dr. LaPerriere himself. For LaPerriere, a retired dermatologist, the museum unites his two passions, medicine and history. "I hated history in school," he said. "It wasn't taught in a very interesting manner, with memorizing dates and all." Medical history stoked his interest, and he's chaired the Medical Society's historical committee for forty years. A visit to the Sacramento City Cemetery at 10th and Broadway twenty-eight years ago drew him to the County Historical Society. The cemetery was in a state of neglect, and that bothered LaPerriere. "I knew what people went through to get here," he said. To honor them, LaPerriere got involved in the effort to restore the cemetery.

The Medical Society had "a lot of stuff sitting in boxes,"

LaPerriere said. To share it with the public, in 1990 he put together a major exhibition for the Sacramento History Museum, called "Out of the Doctor's Bag." The exhibit was a success, and formed the basis for the current medical museum which opened when the Society closed its library and space became available. (The museum still has a large number of historic medical texts. LaPerriere's favorite is the old journals with advertisements. Who knew there were TB sanatoriums all around California, advertising for clients?)

Today, the museum is still growing and LaPerriere is still digging through boxes. He's been invited to sift through old medical stuff in people's attics and garages and occasionally finds treasures—and hazards, too, such as a bottle of the old anesthetic ether. Had the bottle actually still contained ether or the products of its decay, it would have been explosive!

The museum is free and open to the public. School groups, 4th grade and up, are especially welcome. Grants are available to pay for transportation. The museum hosts a lecture series on topics of interest to the general public, such as California plagues and battlefield medicine. Visit the website ssvms.org/museum.aspx for details, or a virtual tour.

Museum of Medical History
Sierra Sacramento Valley Medical Society
5380 Elvas Ave
Sacramento, CA 95819
916/452-2671
Monday-Friday, 9 AM-4 PM except holidays

Science underpins hobby for local snake enthusiasts

This is another topic I have to thank (or blame) my daughter for. She loves snakes, and wanted to get one as a pet. Problem is, snakes live a long time and her time living under my roof was running short. So instead of adopting, she got involved with NorCal Herps and provided temporary foster care to orphan snakes instead.
(Yes, I know how crazy that sounds.)

~~~~~~~~~~

I knew this was an unusual community science event when the announcer asked the audience, "Does anyone have a reptile on you?"

The Northern California Herpetological Society (NCHS), a group of Sacramentans interested in the conservation, care, and breeding of "herps" (reptiles and amphibians), meets the fourth Friday of every month at the Arden-Dimick Library. The meetings are open to the public and though I'm not a snake owner, I was intrigued. I'd heard that UC Davis PhD student Donnelly West was slated to discuss the genetics of color and pattern in ball pythons—with live examples.

And so I met Peter the Piebald Python and discovered people committed to learning the latest herpetology science for the benefit of their zero-legged friends. According to NCHS President Linda Boyko, "Science is a very big part of NCHS." At the time of NCHS's first meeting in 1982, technical information about health and genetics for captive snakes and lizards was hard to come by. Long-time member Sue Solomon says, "Herpetoculturists generally caught and kept wild snakes because there were very few captive breeders and sellers. As far as I know, nobody had lizards except for chuckwallas, which were relatively easy to capture in southern California and Mexico, and iguanas in Mexico and Arizona." (Nowadays, to conserve wild populations, NCHS promotes the acquisition of captive-bred animals.) In those pre-Internet days, NCHS was formed to help owners learn from one another and from experts.

Experts, many of them from UC Davis, give lectures at NCHS every month. Recent topics include the ecology of California's red-legged frogs, and fungal diseases of snakes. The day I attended, I got an excellent overview of dominant and recessive genes and learned that just as some people breed dogs or horses for unusual coat colors, some breed snakes.

If the thought of breeding snakes for fun freaks you out, NCHS members argue that the best way to overcome a fear of snakes is to learn more about them. Alexandria Fulton, Events and Programs Director for

NCHS, says, "Most of the 'scary' behavior of a snake is just them trying to protect their lives despite their small size and the rather large size of a human."

Is it true that if you meet a snake in person you'll find they aren't as intimidating as you thought? I confess that I was surprisingly charmed by West's ball pythons. These medium-sized, mellow reptiles traveled in a big plastic tub, each snake inside its own pillowcase-like bag, where they seemed perfectly content to ball up and keep warm. When lifted out to meet the crowd, each python had its own personality, wrapping around its handler's arm and raising its head to say hello.

(I expect I have failed to convince the skeptics, but believe it or not, they were actually cute.)

Reptiles and amphibians besides snakes are also part of the club. At the meeting I attended, a bearded dragon (a type of lizard) clung to its owner's shoulder. NCHS members also work with the Sacramento Zoo on a nationwide citizen science project called Frog Watch. In this program, people monitor the frog calls from a local water source and record the information in a national database. This helps scientists monitor amphibian populations, which are considered a marker of environmental health.

NCHS is committed to rescue and rehabilitation. Members provide foster care and adoptions of abandoned animals. Boyko says, "We rescue many discarded reptiles that come to us or are found outdoors despite being non-native species. It's sad that many think it's OK to just put your pet out with the trash." Released pet herps usually perish, but even worse, non-native animals might establish a breeding population in the wild. The red-eared slider turtle and the common watersnake are examples of invasive species that threaten California habitats as a result of people releasing their pets.

Education is another club priority. NCHS strives to provide quality live animal education presentations to the public through schools, youth and civic groups. They participate in many public animal outreach events. Members use live animals to teach about care and husbandry procedures for reptiles and amphibians.

If you're intrigued, NCHS invites you to attend a monthly meeting. Visit norcalherp.com or Facebook norcalherp for the latest news. The website also shows animals available for adoption, offers animal care information, and lists local veterinarians who work with herps. To bring Donnelly West's friendly ball pythons to your party or educational event, visit papayapythons.com or Peter Pied Python on Facebook.

Meetings open to the public
4th Friday of the month, 7 PM
Arden-Dimick Library
891 Watt Ave Sacramento, CA 95864

High school students encounter science and medicine in the neighborhood

Cristo Rey Sacramento, "the school that works," is part of a national network of Catholic college preparatory high schools exclusively for students with limited financial resources. Some of these kids face barriers that would stop you in your tracks. Through Cristo Rey, they not only get a first-class private education, they also gain real-world work experience one day a week. Some of that work is STEM related.

Alex Morales can tell at a glance which piece of human tissue is from a smoker's lung. And he hardly even notices the weird smell anymore.

Morales is not your average high school freshman.

Thanks to an innovative Work-Study program at Cristo Rey High School, scores of Sacramento teenagers are getting real-world experience with science and medicine. Rather than just sitting in a classroom reading about biology, these students are also participating and contributing on the job in laboratories, hospitals, and clinics.

According to Sister Eileen Enright, president of Cristo Rey, "The kids really love going into the hospital setting." This year {2015}, seventy-nine of them are working in health care through Dignity and Sutter health systems, UC Medical Center, Kaiser Hospital and several medical and dental offices. "One of our students watched a kidney transplant surgery…This is night and day different from being in class."

One day each week, Morales trades his school uniform (black pants, purple dress shirt, and necktie) for hospital scrubs and reports for duty at Mercy San Juan Medical Center in Carmichael. There he assists in the pathology lab, taking notes and helping with his trusty clipboard while the pathologists process bits and pieces of the human body. "I get a front view of how they analyze specimens," Morales says, noting that much of the work he sees involves fixing tissue so it can be sliced into thin sections to view under a microscope. "This job gives me good work experience in a hospital." And it gives him a perspective on his accelerated math and science courses that is unique in a kid his age.

Edgar Pintor sees the whole person, not just a surgically removed part, in his job with the neuroscience center at Sutter Medical Center. Helping deliver food and water to patients in the hospital, restocking supplies, and visiting with people who frequently are in pain is rewarding for this Cristo Rey senior. "I try to calm their nerves," Pintor says of the patients who may be preparing for surgery on their back or neck. "Sometimes, they're grumpy but after my visit they cheer up and thank me."

Pintor's classmate Karla Davila reports a similar experience in her work at Mercy General, where she spends most of her time with patients who have recently had heart surgery. "I talk to them about coming to our exercise classes," she says. Of the mostly elderly patients, Davila says, "Some are receptive and some don't want to talk." Using her bilingual skills in Spanish and English is a tool she has to help her connect with some patients. Seeing "how physical therapy helps them continue with their daily lives" has inspired Davila to "work even harder to pursue a career in a medical field."

Observing MRI scans and even an esophageal surgery has been part of junior Frida Sarabia's experience working for the cancer navigators at San Juan. According to Kristie Pellerin, who teaches chemistry and anatomy/physiology at Cristo Rey, these kinds of eye-opening adventures improve student learning in the classroom. "They're getting to see and hear a lot. One of my students really made the connection when we introduced 'bicuspid valve,' a term he'd heard at work."

Work experience to the tune of 1,500 hours by the time Cristo Rey students graduate is a cornerstone of the high school. In order to provide a private, college preparatory education to kids who come from poverty (a requirement for admission to Cristo Rey), the school relies on businesses and corporate partners to sponsor the students, allowing the students to offset much of their tuition.

Encouraging interest in science, technology, engineering, and math (STEM) careers for these kids is an explicit goal for David Novak, an engineer by training who's taught physics and math at Cristo Rey for five years after spending 36 years at Jesuit High. "We've started the area's only high school chapter of the Society of Hispanic Engineers {in partnership with a chapter at CSUS}." On project days, Novak guides the students in building and testing "something that moves," including kites, rockets, and mousetrap-powered cars. Celebrating Pi Day, Rubik's cube tournaments, and the lunar eclipse are also part of the club's outreach to inspire STEM ways of thinking.

But there's nothing quite like hands-on science and medicine in the neighborhood to spark enthusiasm for STEM careers. Pellerin observes, "The students who work in the medical field are more interested."

If your workplace could use the competent help of a motivated Cristo Rey student, please contact Leroy Tripette **ltripette@crhss.org** *about sponsoring a Work-Study position.*

Time to rediscover the Discovery Museum

If all goes well, this article will quickly be out of date. The Powerhouse Science Center is on the verge of big changes and great things, if the money comes through. Stay tuned, or better yet, be a part of the change with your support!

I've been a member of Sacramento's Discovery Museum, a hands-on science center on Auburn Boulevard, since the early 2000s when my kids were young. The modest facility has always done its best with what it has. We loved to visit the animals, the planetarium, and especially to participate in the Challenger missions. But Discovery is a small-city attraction.

The times are changing!

After years of difficult starts and stops, the Discovery Museum is transforming into a world-class community science center and educational venue. The transformation in name has already happened: Discovery is now the Powerhouse Science Center-Discovery Campus. The mother ship, Powerhouse Science Center, will be built on the Sacramento riverfront. Powerhouse gets its name from its future main building: the PG&E power station, which is listed on the National Register of Historic Places.

It's going to be spectacular, and visible to all those Bay Area folks driving home from Tahoe.

Preserving and renovating the historic structure is a priority for the City of Sacramento, and Mayor Steinberg is a strong advocate for rejuvenating the entire river and museum district downtown. The time is right. Kaiser will build their new hospital campus just across the freeway from the Powerhouse site. Next door to the south, Sacramento Tree Foundation is spearheading the development of a visionary public park space with cherry trees, dubbed The Hanami Line.

Shahnaz Van Deventer, Director of Marketing and Development for Powerhouse, is bursting with excitement about what's to come. "Thanks to our founding partners, educational grants, and the city council under Mayor Steinberg, we may finally have an official groundbreaking soon." Van Deventer shared with me some of the wonderful exhibit ideas being planned for Powerhouse. Nature is an obvious entry point for kids to encounter science, so Powerhouse will have a nature detectives theme to inspire future citizen-scientists. Keeping it local, the story of water in California will be another featured exhibit, and well as one on energy. Health and bioscience are anchors of our local economy, and Powerhouse will bring them to museum visitors. Space, a perennial favorite theme at the Discovery campus, will travel to Powerhouse with a new emphasis on what's exciting to kids today—Elon Musk, anyone? A planetarium is in the plans, and plenty of space for temporary exhibits that will bring the best of science education on the national scene to Sacramento.

Van Deventer is personally most enthusiastic about the potential to include a space devoted to "design thinking." "This is what's missing from schools," she says. "The chance to experiment and to *fail.*" A design lab could allow hands-on activities where kids discover that failure directs innovation, and multiple iterations of an idea ultimately can lead to success.

In the meantime, the Discovery campus continues to offer its nature discovery, planetarium shows, regular exhibits, and weekend special events like the Big Build Bonanza, when guests apply their own ideas to materials provided by the museum to engineer—well, anything! Summer camps and special workshops for homeschoolers are also on the menu.

More than ever, the new Powerhouse will emphasize community education. Though thousands of people will visit the waterfront facility, many more students need what Powerhouse has to offer. Outreach programs and mobile education to take the museum into schools will be a priority.

Visit powerhousesc.org for the latest news, or to support this exciting new science venue in Sacramento.

Part Six

Health

The not-so-shocking truth about AEDs

I see AEDs hanging on the wall in all kinds of places but I didn't think that had anything to do with me. Wrong: they're not reserved for medical professionals. They're for me, and you, to use in an emergency. No qualifications or training required, just compassion and quick thinking. Here's how.

You're at the grocery store, or maybe the airport. You see a person collapse. On the floor, the person is not moving, not speaking, and you can't wake them up. This person may be one of more than 350,000 Americans who experience sudden cardiac arrest outside a hospital every year. Their heart has stopped pumping oxygen-rich blood to their brains and bodies. In a few minutes, they will die. What do you do?

The top priority is always to call 911 and get emergency services on the way. In the meantime, however, time is of the essence. The victim's chance of survival decreases about 10% with each passing minute of inaction. A bystander using an automated external defibrillator (AED) plus chest compressions can mean the difference between life and death.

A bystander? You mean, *me*? I've seen AEDs hanging on the wall but I can't use one, can I?

Yes, you can. According to Chris Harvey, Public Information Officer for the Sacramento Fire Department, "It's very important not to hesitate. The whole point is to work quickly."

AEDs might look intimidating but they're designed to walk you through each step, even if you have no experience. Just turn on the AED and the device will literally tell you what to do with voice prompts. It will direct you how and where to hook up the sensor pads on the patient. It will instruct you when to perform CPR. Even if you haven't been trained in CPR, you can—and should—do chest compressions when told to do so.

The AED also does the thinking for you. It will decide whether the patient is in cardiac arrest, and if a shock to the heart might help. The question is whether the patient's heart is in a specific kind of abnormal rhythm called ventricular fibrillation, a pattern Harvey describes as the heart moving "in an uncontrolled, unfocused way." If ventricular fibrillation is detected, the AED will charge its capacitor and ask you to make sure no one is touching the patient. When you push the button, an electric shock is delivered to reset the heart's conduction system with the goal of restoring a normal beating rhythm.

Science in the Neighborhood: Health

While that might sound dangerous, Harvey emphasizes that the most important thing to know about an AED is, "It won't shock somebody unless it detects that specific heart rhythm. You can't cause more harm, you can't shock someone who doesn't need it." As intimidating as an AED might be, "Don't be afraid to take it off the wall and use it." (If that assurance isn't enough for you, know that Good Samaritan laws protect you from liability if you act reasonably and in good faith.)

Harvey recommends that we get in the habit of noticing AEDs in places that we frequent. In the building where you work, a theater or restaurant that you visit a lot, pay attention and know if there is an AED and where it's located. More than 2.5 million AEDs have been sold in the US for use by laypersons. Most AEDs are found in public places where people gather such as government buildings and airports, as well as sports clubs, schools, and casinos. Their locations are marked with signs.

Another way to find an AED is to use the PulsePoint app on a smart phone. This app identifies the location of registered AEDs in public places near you. Looking around my neighborhood, the app tells me not only that the nearest AED is at the Bureau of Reclamation building on El Camino, it also describes exactly where in the building it's located and includes a photo of it. Conversely, if you know of an AED which is not yet registered with the app, you can submit details and a photo to get it included.

PulsePoint also is connected with the emergency response systems of Sacramento's Metro Fire District (county) and the Sacramento Fire Department (city). The app "knows" when a possible case of sudden cardiac arrest has been called in, and where the patient is located. This information can be used to get help to the patient even before EMS arrives. How? CPR-trained citizens are invited to register for alerts with the app. If you are in the vicinity of a cardiac arrest, your smart phone will notify you—and tell you where the nearest AED is. Arriving as little as a minute or two before EMS can make a big difference in the patient's chance of survival.

A CPR/AED certification course only takes about 3 hours (2 hours online, 1 hour at the American Red Cross near Cal Expo). Register at RedCross.org.

Sports drink or water?

Store shelves are full of products that make dubious claims about health and wellness. I could write a whole book about the supplement industry, but here I've focused on the single question, are commercial sports drinks better for you than water?

This article ran in January.

~~~~~~

At fitness clubs around the region, new faces are turning up as people resolve to improve their health in the new year. They'll find a plethora of choices. Within the Arden-Arcade area, clubs offer dance fitness (Latin-inspired Zumba and Cardio Dance at California Family Fitness), Les Mills programs (BodyAttack and BodyPump at Del Norte), cycling and power yoga (Arden Hills), and even Spiderman moves on ropes (Bodyweb at Crunch). While sweating and panting through a group exercise class, participants are likely to get thirsty.

During exertion, our bodies lose water. We exhale humid air in our breath, and we sweat to control our body temperature. To prevent dehydration, the lost water must be replaced by drinking.

By drinking what? Water, or a sports drink?

Many people nowadays are choosing commercial sports drinks over old-fashioned water. Beyond marketing, the logic is this. Fluid lost through sweat isn't pure water. It contains salt. Because both water and salt are leaving the body in sweat, it makes sense that a drink which contains both is the way to replenish them.

But is a sports drink really superior to water for hydration?

Sports drinks contain electrolytes. This is a fancy word for dissolved salts. When a grain of table salt (sodium chloride) enters water, it falls apart into separate atoms of sodium and chloride. The sodium and chloride become ions. That is, they carry an electrical charge (sodium positive, chloride negative). Because these ions are charged particles in water, they can carry an electric current—hence the term *electrolytes*.

The electrolytes sodium, potassium, and chloride, along with calcium, magnesium, and others in smaller amounts, are essential for life. Human blood is salty, and the fluid inside cells is loaded with electrolytes. In the body, electrolytes are necessary for fluid balance, muscle contraction, nerve impulses, heartbeats, transporting chemicals into and out of cells, and much more. Regulating the amount of these ions in the blood is the job of the kidneys, which control how much water and electrolytes are dumped in urine and how much are retained.

Electrolytes lost through sweat are replaced in the diet. As anyone who is trying to limit their sodium intake can testify, salt is abundant in the foods we eat. After exercise, normal food plus plenty of tap water will replenish losses without the need for a special rehydrating beverage. For example, Gatorade is a solution of sodium, potassium, and phosphate salts with sugar, food coloring, and artificial flavors added. A twelve-ounce bottle contains 160 milligrams of sodium. That's about the amount of sodium in an equal volume of milk, or a cup of raisin bran cereal, or maybe ten dry roasted peanuts. Twelve ounces of Gatorade provides 45 milligrams of potassium. An average banana packs over 420.

So after a workout, water plus a snack is just as good—or better—than Gatorade for replacing electrolytes. Is there any reason to consume a sports drink *during* exercise?

Yes, but only during vigorous, prolonged activity. As a general guideline, sports drinks have no advantage over plain water during the first hour of exercise. As long as you're drinking water and have healthy kidneys, your body can manage quite a bit of sodium loss. Even among marathon runners, hyponatremia (the condition of having too little salt in the blood) is rare. If you're doing one group class at the gym, water is all you need, and you can avoid the extra calories found in sports drinks.

For athletes who are going harder and longer, or who are exercising in hot, dry conditions that promote a lot of sweating, sports drinks can be useful. In addition to replacing electrolytes (salt), sports drinks include sugar (carbohydrate) in the mix. Sugar water gives a jolt of easily absorbed energy. As a bonus, sugar also helps with hydration. One important transport protein located in the lining of the small intestine only absorbs salt when sugar is present. The protein systematically grabs one glucose molecule and one sodium ion from the gut and moves both of them simultaneously into the circulation. Water follows along. Thus, adding sugar to an electrolyte drink helps the body to absorb salt and water.

(Understanding this co-transport phenomenon has saved countless lives. Victims of life-threatening diarrhea—primarily children in developing countries—can be treated with oral rehydration therapy, a solution of both salt and sugar somewhat similar to a sports drink.)

The sugar benefit can be overdone. If the concentration of sugar in a drink is more than about 8%, it slows the rate at which fluid leaves the stomach. This impairs rehydration. So high-sugar beverages like soft drinks and fruit juices are not a good choice for hydration during a workout. (Plus, they have few electrolytes.)

The most important part of staying hydrated during exercise is the "hydra-" part: water. Whether you choose to fill your bottle at the tap for free, or pay several dollars for a colorful, brand name sugar-salt solution at the store, keep drinking to stay in peak form.

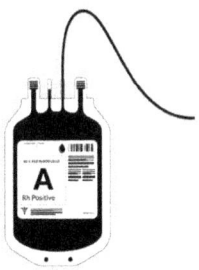

Blood donation and banking

My PhD is in immunology so I was excited to visit a blood bank, and talk about antibodies and such with the experts. The ability to match blood types in order to safely transfuse a patient is a marvel of modern medicine. Learn something about it here.

The first part of this article ran in October

Blood donation

In a quiet room at the corner of Morse and Fair Oaks Boulevard, ordinary people sip juice and munch on cookies.

For this snack, they paid in blood.

This month of the year when vampires and other bloodthirsty creatures of legend get so much attention, it's good to be reminded that human blood really is a magic potion. According to Alexander Sigua of Blood-Source, our region's largest blood bank, every two seconds someone in the US needs a transfusion.

Blood and medicines made from blood save lives. There is no artificial substitute for human blood. Scientists have experimented with "growing" human blood in genetically engineered pigs, but for the foreseeable future, the only source of this vital substance is the generous people who are blood donors. While donors do earn "points" that can be redeemed for prizes, blood donation in Sacramento is just that—a donation freely given with nothing in return.

So far this year, more than 2,500 of these everyday heroes have passed through the doors of BloodSource's Sierra Oaks donor center. Most healthy people 16 years or older, who weigh at least 110 pounds and aren't pregnant, are eligible to donate blood. The majority of donations come from committed, repeat donors, but because the need for blood products is ongoing, recruiting new donors is always important. BloodSource works with the community to run mobile blood drives that reach out to new donors. Alexander Sigua says, "A lot of first-time donors came out for the Sacramento Kings blood drive. They wanted to donate on the court at Sleep Train Arena—and they got free tickets to a Kings game." Another successful event is the annual Causeway Classic blood drive, a competition between UC-Davis and CSU-Sacramento. Public-spirited Aggies and Hornets compete for a trophy awarded at the football game of the same name.

Blood collected at the Sierra Oaks location and elsewhere in the Sacramento region is used locally, with much of the donated blood products going to help patients at

Science in the Neighborhood: Health

Mercy, Sutter, and UC-Davis hospitals. In cases of special need, BloodSource shares with other blood banks. Alexander Sigua says, "We were called on 9/11, and after the movie theater shooting in Aurora, Colorado."

Blood banks also collaborate with a kind of library of hard-to-find donation types. For example, a Sacramento donor might have an unusual profile of certain red blood cell proteins. BloodSource will freeze that unique donation and record it in their "catalog." If a patient in another part of the country needs that particular blood product, their blood bank can request it.

Most donations are of whole blood. Drawing a unit of blood (roughly two cups) takes about ten minutes. Normally blood clots outside the body, so the sterile collection bag contains an anticoagulant. Whole blood is a complex mixture of red and white blood cells, a protein-rich liquid called plasma, and platelets, tiny cells whose job is to stop bleeding. Each of these components has a particular medical use. Because not every patient needs red cells, plasma *and* platelets, a single donation can be separated into these parts and distributed to help more than one recipient. Plasma taken from whole blood donations is sent to a special manufacturer to turn it into medicines that treat hemophiliacs and people who don't make enough antibodies.

In Sacramento, BloodSource can use special techniques to take only plasma or only platelets from a donor. Because these blood components are replaced in the body faster than red blood cells are, platelet and plasma donors can donate more frequently. The technique, called *apheresis*, involves drawing blood and spinning, or *centrifuging*, the blood to separate the components by density.

The desired fraction is removed, and the rest of the blood is continuously returned to the donor.

The number one goal at the Sierra Oaks donor center is safety, both for the donor and the recipients of the donor's gift. Safety begins by asking potential donors a lot of questions about their health, where they've traveled, and certain behaviors that affect the risk of unknowingly carrying an infectious disease. Donors are then given a quick physical exam: vital signs are taken, and a finger prick yields a few drops of blood to test for hemoglobin, an iron-containing protein that gives blood its color. One common reason a willing donor might be turned away is if she doesn't have enough hemoglobin to spare. A blood donation could put her health at risk. Often this deficiency can be corrected by increasing the iron in her diet.

Donors are carefully screened to minimize the chance of their blood transmitting an infection, but a lot more needs to be done before their blood is ready to give to a patient. Let's follow a donation to BloodSource's laboratories at Mather Field and learn some cool science.

Blood banking

After blood is collected, where does it go?

Dr. David Unold, a Yale-trained specialist in transfusion medicine, knows. As he led me through the spotless white corridors of BloodSource's laboratories near the VA Hospital at Mather, he explained the blood bank's strict quality control. "This is a tightly regulated industry," he said. "To begin, each donation is tested for the correct volume. We also assure that the numbers of red blood cells, white blood cells, and platelets are in the proper range."

Blood products are whisked to the bank the same day they're donated. For plasma donations, speed is essential because the precious clotting factors in plasma decay with time. Once plasma has been tested, it's frozen and can be stored for up to a year. Platelets, on the other hand, are damaged by the freeze/thaw process and are never frozen. They're stored at room temperature and must be used within a few days.

Whole blood, the source of red blood cells (RBCs), is transported to the bank on ice. At the laboratory, the blood is centrifuged. Like a clothes washer spinning out water, a centrifuge separates the RBCs from the liquid part of blood and most other cells.

The RBC fraction is then passed through a filter. Red cells navigate the tiny pores but larger white blood cells are trapped. This process of *leukoreduction* minimizes the number of white cells that sneak into a red cell transfusion. While white cells aren't necessarily a problem, they can contribute to fevers in recipients, and can cause long-term issues in people who receive blood frequently. "After leukoreduction, the RBCs go into quarantine," Dr. Unold said. How long? "24-48 hours while we do tests."

Crucial tests characterize the blood type, and screen for blood-borne infections.

During the quarantine period, donors might fall ill with a cold or flu. They're asked to call the blood bank if this happens. Unfortunately, in the case of some dangerous infections, donors may not *feel* sick, so all donated blood must be tested. According to Dr. Unold, Blood-Source screens for viruses, including human immunodeficiency virus (HIV/AIDS), hepatitis B and C, HTLV (a cancer-causing virus), and West Nile. In addition, they test for the sexually transmitted bacteria of syphilis, and in first-time donors only, the parasite that causes Chagas disease (found in South America).

How can they tell blood may be contaminated? "We use both nucleic acid and antibody testing." Nucleic acid testing directly looks for the DNA or RNA of a virus. Ultra-sensitive techniques can amplify and detect the tiniest quantity of a virus's genetic material in the blood.

Antibody testing is indirect. It looks for evidence of the body's response to a viral infection, not the virus itself. If a donor was exposed to a virus, their immune system will have produced antibodies against it. If the blood contains antibodies against a virus, then you know the virus is either in the blood now, or was sometime in the past.

Transfusions of red blood cells and some blood products must be matched according to blood type. The major ABO blood types refer to two sugars, A and B, that can be found on the surface of blood cells. A person's genes determine whether they have the A sugar (type A), the B sugar (type B), both A and B (type AB), or neither (type O). The sugars themselves are harmless, but they act as *antigens*, or targets for an immune response, in people whose bodies are not familiar with them. For example, if a type O person receives blood with the A sugar on it, antibodies in the person's body will immediately attack the transfused "A" blood and cause a massive, even deadly, reaction.

Matching the ABO group (plus the Rh antigen) is all that's needed for a safe transfusion in most recipients. However, a plethora of minor blood antigens do exist. Some patients, especially people who've had transfusions before, have an immune reaction against these less-potent antigens. To help physicians with complex cases like this, BloodSource has a specialized laboratory that can make highly detailed matches between a particular patient and a particular unit of blood.

Some donors have unique blood profiles that make their red cells especially valuable. "Normally we only keep RBCs for a month, refrigerated, but rare units we can freeze for up to ten years," Dr. Unold said.

In the end, every donation is needed. Even units that reach their expiration can be used for certain clinical assays and for research. How willing are Sacramentans to open a vein to help others? According to Dr. Unold, "We have a lot of great donors who make sure we have an ample supply of blood for everyone." Nevertheless, new donors are always needed. Find a center or blood drive near you at bloodsource.org.

Donating "good" bacteria can save lives

> My first love in science is microbiology. I've taught it, I write thriller novels about it, and new discoveries in this field continually change the way I see the world. Microbiome research has certainly done that. We've learned that microorganisms are everywhere—on us, inside us—and they're *not* our enemy. In fact, some of the most unlikely ones are a powerful new medical therapy...

A Sacramento company has an unusual request: Come poop with us.

AdvancingBio is a not-for-profit stool bank that lifted its (toilet) lid in February 2015. Yes, a stool bank. A donation and collection center like a blood bank, except…well, you know.

I am not making this up.

Manager Erin Stinson showed me around the

Science in the Neighborhood: Health

AdvancingBio facility at Mather. My tour began with the donation restroom (very clean and ordinary) and ended in a modest laboratory where the world's most unwanted material is converted into a miracle cure.

In spite of—or in a way, because of—modern medical technology, the hottest new therapy around is human feces. Raw, untreated, brown gold has the potential to save tens of thousands of lives a year in the US, and end misery and debility for hundreds of thousands more.

I'm talking about fecal microbiota transplantation, or FMT, an FDA-approved experimental treatment for recurrent *Clostridium difficile* colitis, commonly known as C. diff. FMT is the infusion of a healthy person's microbe-laden stool into the diseased colon of a person whose own microbiome (gut bacteria) is completely out of whack.

If that sounds gross, Stinson assures me that C. diff infection is a lot worse. "These patients suffer so much," she says. "If I had a way to end that kind of physical pain, I know I'd want it."

C. diff is an infection usually acquired during a hospital stay, and both the rate of infection and the severity of the disease have increased markedly since 2000. Hundreds of thousands of cases are reported, and tens of thousands of people die every year. Ironically, antibiotic treatment is the number one cause of C. diff. Drugs that can save a patient's life by killing bacteria end up triggering a new problem.

The reason for this unintended consequence is the microbiome. Your body, especially the inside of your gut, is loaded with bacteria. Though this fact is under-appreciated, having a wide variety of microorganisms living on and in you is both normal and desirable. Your microbiome is essential for your health in myriad ways that science is just beginning to understand.

One benefit of a healthy gut microbiome is to keep dangerous, disease-causing bacteria at bay. This is called "colonization resistance" and can be likened to the way a forest of tall trees minimizes undergrowth. If mature trees (bacteria of a healthy microbiome) are destroyed by fire (antibiotics), a bunch of different plants (C. diff) can quickly shoot up.

The primary treatment for C. diff is *more* antibiotics. Sometimes that works; C. diff gets knocked out and the microbiome reasserts itself. Other times, the microbiome is too badly damaged to recover, and C. diff just keeps coming back. (*Clostridium* is a particularly tough bug to get rid of because it can enter a dormant form called a *spore* which is impervious to drugs.)

Transplanting healthy stool with its diverse populations of microbial life can restore the gut microbiome, driving C. diff bacteria back into the shadows. In fact, the cure rate for recurrent C. diff after FMT is about 90%. In medicine, that's an unheard-of success. Demand for FMT is growing exponentially. But where to get the poop? And how to guarantee that it's safe?

A few years ago, a group of Sacramento gastroenterologists contacted the CEO of BloodSource, our local blood bank, and said, our patients need this. What can you do?

From that meeting came AdvancingBio, now one of only two non-hospital stool banks in the country. AdvancingBio screens donors with detailed medical questionnaires and blood tests to assure the safety of the donations. Stool is tested for known germs. Then it's frozen in liquid form with a preservative, and shipped to physicians on request. Donations are *not* pooled together, and there isn't any screening for "good" bacteria because we know too little about what to look for. Presumably, whatever microbiome a healthy person has is a good microbiome.

At the donation center, Erin Stinson showed me the complimentary coffee and prunes. I pointed out that everybody poops, no needles required, so getting donations should be easy. Wrong, she said. For one thing, unlike blood donations, "you can't schedule stool donations." For another, "Proximity matters." When nature calls, you need to be close by. Stinson has gamely walked the neighborhood around her center to let people know about the need for donors in the vicinity.

Dozens of people have answered the call, including some kind folks from the nearby Sacramento Fire District headquarters. All donors are volunteers, but Stinson says, "We give prizes for the most donations, and the largest donation." (Yes, size matters.) Repeat donors calling themselves names like Poop Tart and Lil' Stinker are listed on a wall-hanging awards chart in the interest of friendly competition.

AdvancingBio ships 40-50 "products" a month, most of which are administered to patients by colonoscopy. They're starting to get more orders than they can fill. Only half-joking, Stinson says, "We need a PoopMobile." I asked why people couldn't do their business at home and bring in donations. She replied that this creates "chain of custody" problems. Donors must be screened, and donors must be matched to their donation.

AdvancingBio only works with physicians. Patients interested in this therapy must bring it up with their gastroenterologist, not approach the bank directly. Can patients get a directed donation from a relative or friend? Stinson says it's possible but not convenient, and also not recommended. Believe it or not, their experience with blood donations reveals that directed donations on the whole tend to be less safe than anonymous ones, probably because friends and relatives underreport their own health problems.

"We're helping a lot of patients who are suffering," Stinson says.

UPDATE: *AdvancingBio ceased operations in the winter of 2017-18. BloodSource, the parent company, is sticking with its expertise in blood banking. The logistics and clinical indications for FMT are still being worked out.*

If you're fascinated by the microbiome, I recommend the book
I Contain Multitudes *by Ed Yong.*
Sacramento readers can also ask me *about my general-interest lecture on this topic.*

Seasonal allergies

I'd heard it said that Sacramento is one of the worst places in America for seasonal allergies. I learned that's objectively not true, but such facts are no comfort if you're one of the sufferers. Here's some information about the allergic immune response, and tips on how to minimize your hay fever.

This article ran in April

~~~~~~~~~~~~~~~~~~~~~~~~~~~~~~

Runny nose, red eyes, itchy skin, and sneezing? Welcome to spring in Sacramento.

Seasonal allergic rhinitis, also known as hay fever or seasonal allergies, affects tens of millions of Americans. If you're one of them, it's no comfort to know that despite local rumor, Sacramento ranks only #88 in the top 100 "most challenging places to live with spring allergies" (Asthma & Allergy Foundation of America,

2014). For Sacramentans with allergies, spring can be a difficult time. The flowers and trees coming to life are lovely to look at but are a visible sign of an invisible menace: pollen.

Although it's weird to think about, plants reproduce sexually. That means there are male and female plants, or parts on the same plant. Those parts produce the botanical equivalent of sperm and eggs that must come together to make a seed. Pollen is plant sperm, a tiny, tough package for the male DNA.

In unlucky *atopic* humans the immune system reacts to pollen by making a particular type of antibody called IgE. Nobody really knows what IgE is good for. It might be useful in fighting parasitic infections. But in hypersensitivity (allergic) reactions, IgE is like an alarm going off in your body. The antibodies trigger a set of responses that manifest in the nose, lungs, throat, sinuses, ears, or skin as the symptoms of allergy. The major trouble sparked by IgE is the release of the chemical histamine from white blood cells called mast cells. Histamine causes itching, sneezing, redness and other nuisances when it binds to cells in affected tissues.

Sacramento's main allergy season is March through June, though with climate change it's creeping into February. The major culprits are pollens from oak, willow, and walnut trees, and a variety of grasses. (Pollen from weeds, such as ragweed, is more of a problem in the fall.) Although flowers produce pollen that you can sometimes see on the flower, flower pollen isn't a big contributor to seasonal allergies because of the way it's dispersed. Flower pollen is sticky. Flowers rely on bees and other insects to transport it on their bodies. Pollen carried by bugs has little chance of entering your nose. On the other hand, plants that we don't think of as flowering—those trees and grasses—are wind pollinators. They throw vast numbers of pollen grains into the air. Some of those grains land in the eyes, skin, and lungs of people.

Allergies (properly called seasonal allergic rhinitis to distinguish from food allergies) and asthma are related. Both involve self-destructive activity from the immune system. But while allergies are temporary and have a specific trigger, asthma is a chronic, long-term inflammation. Asthma is associated with air pollution and air quality, though it can also be worsened by pollen. Allergies mostly affect the eyes, nose, and skin. Asthma is a disease of the lungs.

If you're bothered by seasonal allergies, there are things you can do to minimize your discomfort. Pay attention to pollen counts. In many locations, the number of pollen grains per cubic meter of air, and mold spore levels, are measured regularly by the National Allergy Bureau, a section of the American Academy of Allergy, Asthma, and Immunology. You can access these counts, and get forecasts of whether they're rising or falling, at many weather forecasting sites and also at pollen.com. If you know what pollen types you're allergic to, you can stay indoors on days when those counts are high.

Don't know which pollens activate your IgE? You might want skin testing done by a physician. Like all antibody-mediated immune reactions, allergies are *specific*. Individuals are allergic to the pollen of certain plants, not all pollen in general. Skin testing can be used to diagnose which ones. A tiny amount of pollen is pricked into the skin. If you have IgE against that plant pollen, a red bump will form.

Pollen counts vary with the time of day and the weather. They tend to be highest early in the morning and on warm, breezy days. On cool, wet days there's generally less pollen in the air. Plan your outdoor activities accordingly. If you've been outside, wash your hands and face, and change your clothes, to prevent pollen getting in your eyes and nose.

Your pharmacy can help you feel better. Eye drops and antihistamines are cheap, effective medicines to treat the symptoms of allergies. As the name suggests, antihistamines work by blocking the binding of histamine to its target cells. Diphenhydramine (brand name: Bendryl), a first generation antihistamine, has been around since the 1940s. It's good at relieving allergy symptoms but it has a number of other effects, such as drowsiness. Second-generation antihistamines such as loratidine (brand name: Claritin) are less sedating because unlike the first generation drugs they don't cross into the brain. When Claritin came on the market in 1993 you needed a prescription to get it. Now it's available over-the-counter, a real relief for allergy sufferers.

Curious to see where pollen and allergies are the worst? Visit the Asthma and Allergy Foundation's "allergy capitals" list (aafa.org/page/allergy-capitals.aspx)

The battle against flu never ends

When I taught microbiology at Sac State, influenza virus was one of my favorite topics. This microbe has it all: history, controversy, ongoing threat, and some fascinating biology. I ended up writing a novel (*The Han Agent*) about a genetically altered flu that targets people of a particular ethnic group. Flu was also an obvious choice for one of my articles, which you'll find below.

For the influenza virus (better known as *flu*), winter is open season on humans. In Sacramento and the rest of the northern hemisphere, that's October to May with a peak in January and February. Flu symptoms (fever, cough, sore throat, runny nose, aching body and head, exhaustion) send a lot of people to their beds, and too many to their deaths.

The reason for the season is only partly understood.

The flu virus drifts through cold, dry air and gets inhaled much better than through warm, humid air. Another explanation for why flu is seasonal: we humans spend more time inside with each other in the winter, creating opportunities for the virus to spread.

Flu isn't just a miserable inconvenience. Influenza *kills* thousands of Americans every year, from a low of 3,000 to a high of 49,000 deaths per year between 1976 and 2006.

How can you protect yourself from this nasty little germ?

There's no shortage of imaginative folk remedies and preventatives for colds and flu. Plenty of them are sold as products in stores, even if there is no evidence that they actually work. Antibiotics definitely do NOT work against influenza, so don't ask your doctor to prescribe them. (The *antiviral* drug Tamiflu can reduce the severity and duration of the flu if you start taking it soon enough.) There are two things, however, that have been proven to reduce your chances of catching the flu.

First, wash or sanitize your hands frequently. Influenza virus lingers on surfaces and is easy to catch when you get it on your hands because we unwittingly touch our faces a lot. (Try to avoid doing that, too.) If someone in your home has the flu, wearing surgical masks can help. A mask blocks the spray of virus-laden droplets into the air when a sick person coughs or sneezes, and helps to keep them from being inhaled by a healthy person. However, some of the particles are too small to be trapped by a mask, so it's only partially effective.

Second, get vaccinated. The CDC recommends annual flu vaccines for everyone over six months of age.

Science in the Neighborhood: Health

You don't have to visit a doctor to get a flu shot. Raley's, Bel Air, Safeway, and pharmacy chains like Walgreens offer it, too. If you hate needles and are between the ages of 2 and 49, you can ask your doctor for a nasal spray of the vaccine instead, though it might not be as effective as the injection. Sacramento County offers free vaccines at flu shot clinics held at various locations in October and November each year. But it's not too late to be vaccinated in December and January. While the vaccine takes about two weeks to kick in, we may have flu in our community into the spring.

So why do we need a flu shot every year, when some other vaccines (such as the tetanus shot) are only required once every ten years or longer?

Influenza is a tricky foe. The virus changes (mutates) at a high rate. Because a vaccine trains your immune system to recognize and attack a very specific target, a change in the way the virus looks can allow it to evade the immune response. Like a master spy, influenza virus puts on a new disguise every year, so every year the immune system has to learn all over again how to identify it.

This phenomenon of flu changing a little bit every year is called *antigenic drift*. It's the reason why last year's flu shot won't fully protect you this year. It's also the reason why some years the flu shot isn't as effective as it should be. Long before flu season starts, scientists predict which versions of the virus are most likely to circulate later in the year. The top three or four candidates are chosen for the annual vaccine. Sometimes a flu virus that wasn't on the radar rises to prominence at the last minute, beating the system. Fortunately, people who faithfully get the vaccine every year develop broader immunity that can help protect them against flu viruses they haven't encountered before.

That wide range of flu immunity might also help against one of the most fearsome killers of all—pandemic flu. Using the analogy of a disguise, imagine that influenza has the unusual ability to not just dye its hair or put on sunglasses, but to have a face transplant. When a flu virus undergoes this kind of sudden and dramatic change in appearance, it's called an *antigenic shift*. Worldwide outbreaks of deadlier-than-usual influenza, such as happened in 1918, can result.

The community's best protection against all influenzas is each of us thinking about others. If you have flu symptoms, cover your cough, and stay home.

Part Seven

At Home

Tasty decay: To ripen fruits, let them talk to each other

Have you ever considered that fruit is alive? The DNA in fruit is still active, and some genes are expressed. As you'll see below, fruits can interact with their environment, responding to external signals and influencing their neighbors. You can use this to your advantage in your kitchen at home.

Ah, the glory of summer produce in Sacramento. Fresh local strawberries, peaches, plums, figs, melons, tomatoes, and more fill our farmers markets and grocery stores. Each of these delights has a moment of peak perfection, when the fruit is fully ripe but not yet mushy, or brown, or syrupy. Unfortunately, produce headed for retail is generally picked before it's ripe. Unripe fruit is less delicate, and better suited to the rigors of packing and transport.

How can a fruit lover get around this problem? Of course the best way to get perfectly ripened fruit is to grow it yourself. Letting fruit ripen on the counter will never be as good as harvesting a just-right tomato directly from the plant. If gardening isn't for you, Kerri Williams, produce manager for the Sacramento Natural Foods Co-op, advises people to buy fruit at different stages of ripeness and eat it as it becomes ready. She also suggests that "if you're not going to consume fresh produce within one to three days, don't buy it. Produce is a living organism. It's not meant to hang around."

But let's say you visit the farmers market at Country Club Plaza on Saturday, or downtown under the freeway on Sunday. Your favorite vendor has special nectarines that you want to enjoy all week. The nectarines are firm but not ripe. Can you regulate the speed at which the fruit ripens so that some are ready to eat the next day, others peak a little later, and the rest are still okay toward the end of the week?

You can. Ripening is a series of chemical reactions. The rate of these reactions is affected by how the fruit is stored.

What we call "ripening" is actually just one phase of the decay process that we later call "rot." Fruits are part of the reproductive cycle of plants. They have seeds. This is in contrast to vegetables, which technically are edible plant parts that don't play a role in reproduction—for example, leaves and stems such as spinach or broccoli. (Elementary school kids will gleefully tell you that tomatoes, avocados, even zucchini are "fruits" by this definition.) True vegetables don't ripen. Fruits do because their job is to prepare and then disperse seeds. The chemical changes of ripening make fruits tastier for animals and birds, encouraging them to pick and eat the fruit, and spread the seeds.

Three things happen when fruit ripens: it becomes softer, sweeter, and it changes color. These effects are caused by enzymes that break down big molecules in the plant into something smaller. Pectinase enzymes degrade plant cell walls, making them soft. Amylase enzymes split starch into simple sugars, which taste sweet. Hydrolase enzymes disintegrate green-colored chlorophyll, bringing out more orange and yellow hues.

This is the interesting part: ripening fruits actually communicate with each other and encourage their neighbors to ripen. You may have heard the saying, "One bad apple spoils the bunch." This is why. An overripe apple or banana strong-arms nearby fruits into rapid ripening, and then rot (an extension of the same process) by turning on gene expression of the necessary enzymes.

So how does a banana talk to a peach? By releasing a gas into the air. Almost all the fruits you eat are sensitive to this gas, and most produce it as well. The ripening gas is ethylene, an odorless, colorless, harmless gas naturally produced by plants. (Oddly enough, in the 1920s, ethylene was used as a general anesthetic during surgery.) Nowadays ethylene is used commercially to ripen bananas for sale in grocery stores. Harvested and shipped green, bananas are placed in temperature- and humidity-controlled rooms. Ethylene gas is pumped in to trigger the ripening process, and when the first hint of yellow appears, the bananas are delivered to retail.

Conversely, you can stop ripening by inhibiting the

production of ethylene. Plants need oxygen to make ethylene, so storing fruit in a cold, humid, low-oxygen atmosphere can essentially put the fruit in hibernation. This technique of controlled-atmosphere storage is widely used for apples, which can be preserved unchanged for up to a year after harvest. That's why you can purchase domestic apples year-round, even though the crop itself is seasonal.

You can use ethylene at home to control the ripening of your own fruit. Here's how.

Place the nectarines you want to eat as soon as possible in a paper bag along with a yellow or brown banana. Ripe bananas crank out a lot of ethylene gas, but other ripe fruits would work as well (apple, cantaloupe, peach, fig, plum, etc.). You can further increase ethylene production by damaging the banana with a bruise or cut. (Fruits respond to injury with a desperate attempt to fast-ripen their seeds before they die.)

The nectarines you'd like to ripen a bit later should go in a paper bag by themselves. They will eventually produce their own ethylene and get ripe, but not as fast as those stuck with a screaming banana.

When some of the nectarines are ripe, store them in the refrigerator. Cold slows down chemical reactions, including those in the ripening process. Your ripe nectarines will eventually progress from ripe to rotten in the refrigerator, but more slowly than on the counter. Just make sure you don't have any overripe fruit in the same drawer, as ethylene gas will build up and make everything decay rapidly.

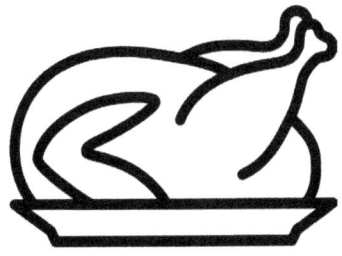

The science of a perfect turkey

Why is it so hard to roast a turkey well?
Science has the answer, and some possible solutions.

This article ran in November

~~~~~~~~~~~~~~~~~~~~~~~~~~~~

Science and cooking go together.

Much of what happens in a chemistry lab resembles cooking. Chemists measure volumes and weights, they mix, heat, and transform one substance into another. Much of what happens in the kitchen is chemistry: salt crystals dissolve, water changes phase from liquid to gas, runny clear egg white stiffens into meringue.

I love the science-y aspects of cooking. I like knowing that olive oil has a lower smoke point than peanut oil, and that enzymes in fresh pineapple (but not canned) will prevent Jell-O from gelling.

So this month, when I'm faced with one of the toughest culinary challenges—cooking a whole turkey that's done, but not dry—I sought advice from Harold McGee's classic book, *On Food and Cooking: The Science of the Kitchen*.

McGee explains everything a kitchen scientist could want to know about cooking meat. He reveals that meat gets juicy at about 140 degrees, when shrinking protein fibers release water. He tells me that denaturation of myoglobin pigment explains the bleaching of fresh red meat when it's cooked, and that the honeycomb structure of bones makes them insulators that slow the transfer of heat.

I'm enjoying all this until I get to a section titled "The Challenge of Whole Birds."

In case you had any doubt, science has proven that roasting the perfect turkey is hard. The problem is, breast meat and leg meat are chemically very different and have different ideal cooking conditions.

In general, turkeys don't fly much. They walk. So in life, the turkey's leg muscles get a lot more exercise than the breast. The more a muscle is worked, the more connective tissue it has. Connective tissue, made mostly of the proteins collagen and elastin, is like a scaffold for the muscle. It provides support for the muscle fibers to pull against. Turkey breast doesn't need much connective tissue. Turkey legs have quite a bit: two to four times more collagen than breast meat.

You can easily tell a high-collagen meat from a low-collagen one by cutting or chewing it. Collagen protein is solid and tough. Meats with little collagen in the muscle are naturally tender. A tough cut of meat can be made tender by cooking. If you heat collagen long enough, it dissolves into gelatin, which is tender and tasty. That's why cheap, fibrous meats are delicious when stewed.

(Incidentally, this is a separate issue from "white" and "dark" meat, which is related not to how *much* a muscle gets used, but the *way* it is used. White muscle fibers are adapted for fast, brief spurts of activity. Red muscle fibers specialize in sustained activity. Ducks, which use their breast muscles to fly for hours at a time, have "red" or dark breast meat.)

So the chef's solution to tough turkey legs should be to cook the meat a long time. Unfortunately, muscle fibers respond to cooking the opposite of connective tissue. Collagen gets softer; muscle gets tougher. (Think of what happens to a low-collagen meat like beef tenderloin when it's overcooked.)

This creates a dilemma for the cook who is preparing a whole turkey. Leg meat needs to be roasted to 165 degrees to get rid of the collagen. But above 155 degrees, breast meat dries out and loses its natural tenderness.

What to do?

One option is to cut the bird up and roast the legs and breasts separately. Another is to try to physically slow down the cooking of the breast. McGee suggests covering the breast with foil, or draping it with strips of pork fat, or before cooking to chill the breast with an ice pack while bringing the rest of the bird to room temperature.

I plan to use a chemical option. Brining can compensate for the tendency of the breast meat to dry out. To brine a turkey, soak it overnight (or longer) in a 3-6% solution of salt water. That's about 2-4 tablespoons of salt per quart; you can add herbs, too.

Salt loosens the protein structure of muscle, tenderizing it, and allowing the fibers to absorb more water. (For you chemists out there, this is an effect of protein-salt interactions, not osmosis, which would do the opposite.) With that extra water on board, brined meat can tolerate some overcooking before it dries out. In the case of a whole turkey, this translates into moister breast meat *and* fully cooked legs.

Brining meat does leave a little salty taste, and the absorbed water dilutes the meat juices, making them less flavorful. But a proper Thanksgiving dinner ought to include other foods to balance this out.

That's a task for the art of the kitchen, not the science.

A blacklight in the darkness

I was searching for a Halloween-themed science topic and got to wondering about glow sticks and blacklights. Learning how these things work was one of the most fun "Science in the Neighborhood" projects for me. As always, my research yielded a practical tip.

This article ran in October

~~~~~~~~~~

'Tis the season for eerie lights. This month you'll see glow-in-the-dark face paint, creepy decorations shining ghostly green under black light, and glow sticks dangling from the necks of trick-or-treaters.

These lights are different from sunlight or ordinary light bulbs. They're low-intensity and viewed best in the dark. They're a single color, and they're cool to the touch.

What are they?

These "glow" lights are all examples of *fluorescence*. Fluorescence is a kind of light produced by a fluorescent molecule (or *fluorophore*) after it is charged with energy. Typically, the energy comes from electromagnetic radiation (EMR)—either visible light or short wavelength, high-energy forms like ultraviolet and X-rays.

When you bombard a fluorophore with electromagnetic radiation (such as by shining a light on it), the fluorescent molecule absorbs the energy but doesn't keep it. Instead, the fluorophore sends energy back out as EMR of a longer wavelength. In other words, it emits light of a different color.

This creates cool visual effects if the "light" used to charge the fluorophore is invisible. Black lights such as you'll find at a Halloween store are an excellent example. Black lights are peculiar light bulbs that emit EMR in ultraviolet wavelengths that are mostly outside the range that the human eye can detect. Even when a black light is burning at full intensity, all we can see is a faint purple glow. But the energy is there, and if it shines on, say, a fluorescent skeleton decoration, the skeleton lights up. Because we can't see the brilliance of the black light, but we can see the re-emitted light coming from the skeleton, the whole thing seems like magic.

But what about glow-in-the-dark T-shirts or watch faces that shine in *total* darkness?

This is another kind of fluorescence that's properly called *phosphorescence*. Phosphorescence is delayed or slow fluorescence. As with fluorescence, phosphorescent substances first have to be activated by exposure to electromagnetic radiation. But instead of immediately emitting energy, they release their light gradually over time.

If you've ever had a glow-in-the-dark item, you've probably experimented with these properties of phosphorescence yourself. To get your item to glow with the highest intensity, you first have to charge it by shining a really bright light on it. The longer you charge it, the more energy it stores, and the longer it will glow later.

A third common example of fluorescence is glow sticks. Glow sticks are a clever way of packaging a fluorophore with a built-in energy source that the user can activate when ready.

As you might guess, the energy comes from a chemical reaction. Inside every glow stick is a brittle, glass-like tube that keeps two chemicals apart. When you bend a glow stick, you break the tube and the chemicals mix. They react, and the reaction releases invisible energy. The energy charges the fluorophore, and the fluorescent molecules glow.

Glow stick light is brightest at the beginning. It fades as the chemicals are used up. You can regulate the reaction rate, and the lifespan of your glow stick, using temperature. Like most chemical reactions, this one is accelerated by heat and slowed by cold. You can't turn off a glow stick, but if you want to save some of the light for the next day, put the stick in a freezer. The reaction will slow dramatically, conserving the chemicals for later. When the stick is warmed again, the reaction will resume and the stick will brighten.

On the other hand, if you want a glow stick to stay illuminated at about the same level for the longest possible time, rather than burning brightly at first and then dimming, refrigerate it before you turn it on. This will slow the initial reaction and even out the light intensity over time.

Note that the fluorophore in a glow stick is not consumed. A glow stick will fluoresce under black light before *and* after it's been used.

Nature invented glow-in-the-dark long before humans turned it into technology. *Bioluminescence* is light from living things. The best-known example of bioluminescence is the firefly. You won't see a firefly in Sacramento, but you can go to the best place on earth to see organisms that light up: the ocean. Many fish and corals make their own light in the darkness using enzymes called *luciferases*.

To witness this marvel, visit Tomales Bay, near Point Reyes National Seashore just north of San Francisco. Tomales Bay is home to billions of tiny bioluminescent creatures called dinoflagellates. Take a nighttime kayak tour at the right time of year and you'll see the waters light up with glowing fairy dust. A number of companies offer these excursions. Summer and early fall are usually the best times to go but these unforgettable tours continue into November.

Solar cooks roast in the sun

I was introduced to solar cooking by my colleague Dr. Robert Metcalf, a microbiology professor at Sac State. Metcalf demonstrated that it's not necessary to boil unsafe water to make it drinkable. A lower Pasteurization temperature is adequate, and a solar cooker can do the job. Using the sun instead of fuel can save lives and time for people in impoverished places. Here at home, solar cooking is "green" and fun.

Some days in Sacramento it feels hot enough to fry an egg on the sidewalk. But it doesn't take triple-digit air temperatures to cook just about anything using California's abundant sunshine. Solar cookers, or sun-powered ovens, concentrate the energy of the sun and provide a free, zero-emissions way to prepare meals even when you're wearing long sleeves.

In our area, solar cooking season generally runs from

April through October. Surprisingly, higher summer air temperatures are *not* the reason. The sun's heat doesn't power a solar cooker. Instead, a solar cooker collects electromagnetic radiation, including visible light, from the sun and focuses it on a black-colored cooking pot. The radiant energy absorbed by the pot is turned into heat which the cooker is designed to trap.

Therefore solar cooking is fastest and easiest not on the hottest days, but days when the sun's energy is at its peak—that is, any cloudless day around the summer solstice, which is in late June. On this first day of summer, the Earth's tilted axis points the northern hemisphere most directly toward the sun, giving us the longest day and the most intense solar fuel for cooking.

What can you cook using the sun? A simple cardboard and aluminum foil panel cooker can cook anything your Crock-Pot can. Stews, either vegetarian or with meat, are foolproof in a solar cooker because it's impossible to overcook. Foods that take a lot of heat on the stove, such as hard-cooked eggs, whole fresh beets, rice, potatoes, and lentils can all be prepared outdoors in a solar cooker. In June and July you can even solar cook baked goods such as banana bread and brownies. When you're running the air conditioner and you dread turning on your kitchen stove or oven, solar cooking can save you money and help to keep you cool.

While solar cooking is a fun hobby in Sacramento, in poorer parts of the world it can change lives. Sacramento-based Solar Cookers International (SCI) is a local nonprofit that works with organizations all over the world to bring solar cooking to those who can benefit from it most, especially poor women and people in refugee camps. After attending a United Nations meeting of the Commission on the Status of Women, SCI's Executive Director Julie Greene said, "Poor women may spend three to five hours per day gathering firewood to cook. On sunny days when they can use a solar cooker it frees up a huge amount of time. Kids can go to school instead of gathering wood."

Solar cookers come in three basic designs. Parabolic cookers look like shiny satellite dishes with a pot suspended in the middle. These cookers have the advantage of getting very, very hot (up to 1000°F) but they're expensive and must be adjusted frequently to follow the sun and keep its rays in focus. Box solar cookers are lined with a shiny, reflective material and have a transparent cover to keep warm air inside. They can be as simple as a pizza box lined with foil, or a more sophisticated version that can reach 400°F.

Panel cookers are the easiest to make and the cheapest to buy. They can be portable and collapsible for easy storage or carrying to a campsite. Panel cookers act like a funnel for sunshine with a set of shiny cardboard panels slanted around a pot. Food is placed in a thin-walled black metal pot with a lid (Granite Ware is ideal). A clear, heat-resistant bag (such as a turkey roasting bag) or an inverted glass bowl goes over the pot to trap heat and moisture. At our latitude in Sacramento, a cardboard panel cooker can reach 220°F.

The intensity of the heat generated surprises people. SCI Director Greene, who is an experienced solar cook, says, "I tell people, don't touch the pot, it's hot! They look straight at me and touch the pot."

You can buy or build a simple solar cooker using plans

on the internet. Solar cooking kits can be an important part of household emergency preparedness, making it possible to cook food and pasteurize water if utilities fail.

But you don't need a crisis to enjoy this special use of solar power. As Greene says, "Solar cooking is fun, it's easy, and it works for anyone who lives where there's sunshine."

Learn more about SCI at solarcookers.org

Swim smart: Six healthy swimming tips that aren't common sense

I began researching this article with the goal of understanding pool chemistry. There is some of that, plus a collection of other swimming-related science tidbits that you can use.

This article ran in July

This month many of us will escape the heat in a pool or the Sacramento or American Rivers. Swim safety is a priority, with the number one goal being to prevent drownings. But there's more to a healthy swim experience than not drowning. To enhance your aquatic adventures, here are six smart swimming tips that come from science.

Clear water isn't necessarily clean water. Swimming

pool water should be clear, but clarity alone doesn't mean the water is free of disease-causing microorganisms because germs are invisible to the eye. A pool's main defense against germs is a mixture of chlorine-containing molecules, including hypochlorous acid and hypochlorite ion (bleach), collectively called free chlorine. At the proper concentration and pH, free chlorine will kill bacteria and other microbes in the water.

If you're wondering whether your favorite pool has the right chemical balance to control germs, you can check it yourself by buying chlorine test strips and dipping one in the water. Or use your nose: there should be a faint whiff of bleach near the water surface.

An excessive "chlorine" smell at the pool means someone needs to add *more* chlorine. Most people think that an eye-irritating odor around a pool is a sign of too much chlorine in the water. In fact, that strong smell isn't free chlorine. It's chloramines. Chloramines are formed when free chlorine reacts with contaminants in the pool, especially urine. The way to get rid of that smell is to shock or superchlorinate the pool with enough free chlorine to turn the chloramines into a gas (ammonia) that dissipates into the air.

Be aware that a stinky pool may also be a germy pool because unlike free chlorine, chloramines are not very good at killing microbes.

Avoid swimming in public waters after a heavy rainfall. Runoff from a storm washes soil, animal waste, and sewage overflow into lakes, rivers, and ocean shores, increasing the number of potentially dangerous bacteria in the water. The effect is temporary; most disease-causing germs naturally die off within a day or two.

Go ahead and swim after eating. Junior just finished a bowl of ice cream and wants to jump in the pool. But Mom puts him on land-locked time out because she knows you shouldn't swim right after you eat.

Nonsense. There has never been a single documented case of a person drowning as a consequence of eating. The swimming-eating myth probably arose from the common sense observation that if you exercise hard on a full stomach, you may get a "stitch" in your side. Few recreational swimmers swim at that level of intensity, and even if you did, a cramp isn't going to make you sink like a rock. Your muscles will work just fine—and get you to the shore or shallow water if necessary.

Use extra sun protection around water. Solar ultraviolet radiation causes sunburns and damages DNA, raising your risk for skin cancer. Being near water increases your sun exposure because some of the radiation is reflected. If you're in the path of that reflected ultraviolet light, you get a double dose, from above *and* below.

The direction of the reflected light depends on the angle at which the sun is hitting the water. At mid-day, when the sun is straight overhead in the sky, light is reflected straight up—at people in or on the water. If you're in a boat, a hat alone won't protect your face. By contrast in late afternoon, the sun is lower on the horizon and light skips off the water at an angle closer to the ground. People on the shore will catch those reflected rays.

Don't borrow air from a scuba diver. Do you ever take a big breath and see how deep you can dive? Or have you been snorkeling and followed a fish to the bottom? Then you've tried freediving, which is swimming deep while holding your breath. Serious freedivers include spear fishermen, and anyone catching California abalone, which by law may not be taken using scuba gear.

Say you're freediving and you encounter a scuba diver. You borrow some air from the scuba tank. This gives you extra time to explore, maybe to bag another abalone, and then you hold your breath and swim to the surface as usual. Sounds great, right?

Wrong. The air from the scuba tank is *compressed* by the pressure of the water column above you. As you ascend, the pressure decreases, and the air you're holding in your lungs expands. Even from a modest depth of say, ten feet, the change in volume of the air can be enough to damage your lungs. If you hold your breath while rising from greater depths, air expansion can kill you.

Index

A

AEDs (defibrillators) 143
Allergies, seasonal 163
American River 26, 29, 30, 33, 35, 55, 83, 113, 189
Ammonia 67
Anaerobic digestion 65, 75, 113
Antibiotics 160
Antigenic drift 169
Atmospheric river 30
Auburn dam 31

B

Bacillus thuringiensis (Bti) 109
Bananas 175
Bark beetles 93
Bioassays 67
Bioluminescence 184
Birding
    Great Backyard Bird Count 97
    Sacramento Audubon Society 98
Blacklight 182
Blood banking 154
    Plasma 155
    Platelets 155
Blood donation 151
Bufferlands 61

C

California Naturalist 115
Cell phones 10
Chlorine 54, 57, 190
Citizen science 34, 94, 98, 116, 130, 138
Climate change 34

Clostridium difficile (C. diff colitis) 159
Coastal Ranges 18, 38, 112
Cosumnes River 85, 97
Cristo Rey High School 133

D

Discovery Museum *see Powerhouse Science Center*
Dopamine 11
Dutch elm disease 93

E

Eames splint 126
Earthquakes 37
    Insurance 39
Eddy current 1, 71, 78
Effie Yeaw Nature Center 116
Electricity
    California Renewable Portfolio Standard 48
    Flex alert 46
    Forecasts 19
    ISO (independent system operator) 44
    Renewables 46, 75, 185
    SMUD 48
Electrolytes 148
Elm trees 92
Emergency vehicles 4
Engineers 73
    Jobs 2, 28, 30, 32, 35, 39, 45, 55, 62, 69, 77
Ethlyene gas 175
E-waste 76
Exercise 147

F

Floods
    1862 26
    1986 26
    Forecasting 32, 33
    I Street gauge 35

Science in the Neighborhood: Index

    Sacramento Area Flood Control Agency 26, 29, 34
        Top of conservation level 34
Fluorophores 182
Fog 21
    Radiation fog 22
    Tule fog 21, 22, 24
Folsom Dam 26, 29, 30, 31, 32, 85
Forman, Bruce 101
Fruit ripening 173

G

Geosynthetic liners 73
Glow sticks 183
Gold mining / Gold rush 27, 32, 84, 111, 112

H

Halloween lights 181
    Glow in the dark 182
Hardpan 89
Northern California Herpetological Society 129

I

Inattention blindness 12
Inductive loop 3
Influenza (flu) 155, 167
Inversion layer 8, 23, 24
Iron lung 126

K

Kitchen science 177

L

Landfill, Kiefer 72
    Landfill gas 75
LaPerriere, Dr. Robert 126
Leukoreduction 155
Levees 26, 27, 29

Liquefaction 39

M

Marine layer 18
Megaflood 25
Mercury 77, 111
Microbiome 158
Mosquitoes 90, 104, 109
    Sacramento-Yolo Mosquito and Vector Control District 105
Mosquitofish 108
Museum of Medical History 125

N

Nature Bowl 101
Nimbus Fish Hatchery 85

O

Ozone 8

P

Pacific Flyway 97
Pacific High 20
Phosphorescence 182
Pineapple express 30, 33
Pollen 164, 166
Powerhouse Science Center 137
PulsePoint 145
Python, ball 131

R

Rancho Seco 49, 90
Recycling 71

S

Sacramento Science Idol 123
Sacramento Splash 88

Science in the Neighborhood: Index

Sacramento Tree Foundation 94, 138
Salmon 56, 83, 86
Sandhill cranes 97
Science cafes 121
Scientist 123
    Jobs 57, 66, 85, 91, 101, 105, 117, 134
Scuba 191
Smeloff valves 127
Smog 8
Solar cooking 185
Sports drinks 147
Steelhead (trout) 84, 85, 114
Stool banking 158
    Transplant / Donation 160
Summer solstice 186
Swimming pools 190

T

Tadpole shrimp 89
Transportation xi
    Distracted driving 10
    Gasoline, summer blend 6
    Traffic lights 1
Turkey meat
    Breast 179
    Collagen 178

U

Underseepage 27, 28
Utilities 41
    Electricity 43
    Household trash 69
    Sewage / Wastewater 59
        EchoWater project 66
        Sacramento Regional County Sanitation District 60
    Water 51
        Sacramento Suburban Water District 52

V

Vaccine, flu 168, 169
Vapor pressure 7
Vernal pools 89

W

Water
    Fresh water
        Surface water (rivers) 55
    Ground water 51
    Wastewater
        Activated sludge microorganisms 64
        Treatment 62
    Wells 53
Water treatment plant 55
Weather 15
    Delta breeze 17
    Flooding 25
    Fog 21
    Forecasting 19, 30, 31, 33, 46
    Marine layer 18
    Pacific High 20
Weir 35, 85

Y

Yellow-billed magpie 98

Z

Zika 105

Did you enjoy *Science in the Neighborhood?*

I would be grateful if you post a review online at Goodreads, amazon, or your favorite social media. Or better yet, recommend the book in person to a friend!

ScienceThrillers Media publishing specializes in page-turning stories, both fiction and popular nonfiction, that have real science, technology, engineering, mathematics, or medicine in the plot.

Visit our website and join the STM mailing list to learn about new releases.

ScienceThrillersMedia.com
publisher@ScienceThrillersMedia.com

Science-themed thriller novels by Amy Rogers:

PETROPLAGUE
Oil-eating bacteria contaminate the fuel supply of Los Angeles and paralyze the city

REVERSION
An American scientist is trapped when a medical tourism center is taken over by a Mexican drug cartel

THE HAN AGENT
A pharmaceutical company with ties to war crimes hires a Japanese-American scientist after she is expelled from a university for mutating flu viruses in the lab.

About the Author

Amy Rogers, MD, PhD, is a Harvard-educated scientist, novelist, journalist, educator, editor, critic, and publisher who specializes in all things science-y. Her novels use real science and medicine to create plausible, frightening scenarios in the style of Michael Crichton. Formerly a microbiology professor at California State University-Sacramento, she is the founder of ScienceThrillers Media and runs the ScienceThrillers.com book review website.

To sign up for Amy's mailing list, click here
or visit her website AmyRogers.com

Twitter: @ScienceThriller
Facebook.com/ScienceThrillers

Invite Amy to speak to your group
amy@AmyRogers.com

www.ingramcontent.com/pod-product-compliance
Lightning Source LLC
Chambersburg PA
CBHW051543020426
42333CB00016B/2068